Lecture Notes
in Business Information Processing 125

W0080331

Jan Mendling
Matthias Weidlich (Eds.)

Business Process Model and Notation

4th International Workshop, BPMN 2012
Vienna, Austria, September 12-13, 2012
Proceedings

 Springer

Volume Editors

Jan Mendling
Vienna University of Economics and Business
Department of Information Systems and Operations
Vienna, Austria
E-mail: jan.mendling@wu.ac.at

Matthias Weidlich
Technion - Israel Institute of Technology
Faculty of Industrial Engineering and Management
Haifa, Israel
E-mail: weidlich@tx.technion.ac.il

ISSN 1865-1348   e-ISSN 1865-1356
ISBN 978-3-642-33154-1   e-ISBN 978-3-642-33155-8
DOI 10.1007/978-3-642-33155-8
Springer Heidelberg Dordrecht London New York

Library of Congress Control Number: 2012945767

ACM Computing Classification (1998): J.1, H.3.5, H.4

Typesetting: Camera-ready by author, data conversion by Scientific Publishing Services, Chennai, India

Printed on acid-free paper

Springer is part of Springer Science+Business Media (www.springer.com)

Preface

The BPMN 2012 workshop series provides a forum for academics and practitioners that share an interest in business process modeling using the Business Process Model and Notation (BPMN), which has seen a huge uptake in both academia and industry. It is seen by many as the de facto standard for business process modeling, as it has become very popular with business analysts, tool vendors, practitioners, and end users. BPMN promises to bridge business and IT, and brings process design and implementation closer together.

BPMN 2012 was the fourth workshop of the series. It took place September 12–13, 2012, at WU Vienna, Austria. This volume contains six contributed research papers that were selected from 22 submissions. There was a thorough reviewing process, with each paper being reviewed by, on average, three Program Committee members. In addition to the contributed papers, these proceedings contain three short papers and one extended abstract of the invited keynote talk. In conjunction with the scientific workshop, a practitioners' event took place on the second day of the workshop. Furthermore, the EMISA 2012 and the AWPN 2012 workshops were co-located.

We want to express our gratitude to all those who made BPMN 2012 possible by generously and voluntarily sharing their knowledge, skills, and time. In particular, we thank the Program Committee members as well as the additional reviewers for devoting their expertise and time to ensure the high quality of the workshop's scientific program through an extensive review process. We are grateful to all the authors who showed their appreciation and support of the workshop by submitting their valuable work to it. Finally, we thank the sponsors of the BPMN-Anwendertag for their generous support, namely, MID, Signavio, Sparx Systems, BPM&O, the City of Vienna, Wirtschaftskammer Wien Unternehmensberatung & Informationstechnologie, and Austrian Airlines.

September 2012

Jan Mendling
Matthias Weidlich

Organization

Program Chairs

Jan Mendling   Wirtschaftsuniversität Wien, Austria
Matthias Weidlich   Technion - Israel Institute of Technology, Israel

Local Organization

Katharina
     Disselbacher-Kollmann   Wirtschaftsuniversität Wien, Austria
Stefanie Errath   Wirtschaftsuniversität Wien, Austria

Program Committee

Thomas Allweyer	University of Applied Sciences Kaiserslautern, Germany
Alistair Barros	Queensland University of Technology, Australia
Marco Brambilla	Politecnico di Milano, Italy
Gero Decker	Signavio GmbH, Germany
Remco Dijkman	Eindhoven University of Technology, The Netherlands
Marlon Dumas	University of Tartu, Estonia
Philip Effinger	University of Tübingen, Germany
Dirk Fahland	Eindhoven University of Technology, The Netherlands
Diogo Ferreira	IST - Technical University of Lisbon, Portugal
Andreas Gadatsch	Hochschule Bonn-Rhein-Sieg, Germany
Denis Gagné	Trisotech Inc., Canada
Felix Garcia	University of Castilla-La Mancha, Spain
Luciano García-Bañuelos	University of Tartu, Estonia
Marta Indulska	The University of Queensland, Australia
Jana Koehler	Hochschule Luzern, Switzerland
Oliver Kopp	IAAS, University of Stuttgart, Germany
Agnes Koschmider	Karlsruhe Institute of Technology, Germany
Frank Michael Kraft	AdaPro GmbH, Germany
Ralf Laue	Westsächsische Hochschule Zwickau, Germany
Henrik Leopold	Humboldt-Universität zu Berlin, Germany
Niels Lohmann	Universität Rostock, Germany
Alexander Luebbe	University of Potsdam, Germany
Bela Mutschler	University of Applied Sciences Ravensburg-Weingarten, Germany
Markus Nüttgens	Universität Hamburg, Germany

Andreas Oberweis	Universität Karlsruhe, Germany
Chun Ouyang	Queensland University of Technology, Australia
Susanne Patig	University of Bern, Switzerland
Artem Polyvyanyy	Queensland University of Technology, Australia
Frank Puhlmann	inubit AG, Germany
Jan Recker	Queensland University of Technology, Australia
Manfred Reichert	University of Ulm, Germany
Hajo A. Reijers	Eindhoven University of Technology, The Netherlands
Stefanie Rinderle-Ma	University of Vienna, Austria
Gregor Scheithauer	Opitz Consulting, Germany
Sergey Smirnov	SAP Research, Germany
Pnina Soffer	University of Haifa, Israel
Lucinéia Heloisa Thom	Federal University of Rio Grande do Sul, Brazil
Hagen Voelzer	IBM Research, Switzerland
Barbara Weber	University of Innsbruck, Austria
Mathias Weske	University of Potsdam, Germany
Stephen White	IBM, United States of America
Karsten Wolf	Universität Rostock, Germany
Peter Wong	Fredhopper B.V., The Netherlands

Additional Reviewers

Eike Bernhard
Olha Danylevych
Niels Mueller-Wickop
Daniel Schleicher
Sebastian Wagner
Michael Werner

Sponsored by

Table of Contents

BPMN Research: What We Know and What We Don't Know

Jan Recker

Professor and Woolworths Chair of Retail Innovation
Information Systems School
Science & Engineering Faculty
j.recker@qut.edu.au

Abstract. In this short keynote paper, I will briefly explore the current state of research and practice surrounding the BPMN standard. On basis of this analysis I will offer a personal outlook into the key emerging areas where I believe more research will be required to further understand BPMN, its premise and promise, and how we can shape – and join together – the landscape of BPMN practice and development in academia and industry.

Keywords: BPMN, process modeling, Known Unknowns, empirical research, design research, research agenda.

1 Introduction

I think we all know by now that the Business Process Model and Notation standard is here to stay. What started of pretty much exactly ten years ago as yet another effort to define yet another modeling approach has over the years become the de facto global standard for the modeling of business processes, or as we call it, the modeling grammar of choice to analyze and design process-aware information systems. The efforts towards BPMN were driven by the ambition to provide a unifying standard notation that could serve all sorts of business users and application purposes [20]:

The primary goal of the BPMN effort was to provide a notation that is readily understandable by all business users, from the business analysts that create the initial drafts of the processes, to the technical developers responsible for implementing the technology that will perform those processes, and finally, to the business people who will manage and monitor those processes.

One may speculate about why BPMN has been successful in its effort to become a widely accepted standard. There are certainly a number of factors that can be attributed to the success: The differentiation of core and extended elements. The provision of advanced and extended modeling capabilities. The promise of model-driven code generation for executable processes. The move to form an OMG working group to gain official recognition of the standard. The realization of both advanced and basic modeling capabilities in a reasonably intuitive format.

J. Mendling and M. Weidlich (Eds.): BPMN 2012, LNBIP 125, pp. 1–7, 2012.
© Springer-Verlag Berlin Heidelberg 2012

In any case, the uptake of BPMN in its version 1.0 was significant, and truly global. By 2008, only one year after the official ratification as an OMG standard was finalized, BPMN was actively used in over thirty countries across all continents [10]. By now, these figures would have increased further. BPMN is available in version 2.0 since 2010 [7], and its application in industry as well as in academia is alive and well. And given the ten-year history, it is only benefitting to pause and reflect about BPMN and its role and prominence in academic research. I welcome the opportunity to share my thoughts on BPMN in academic research.

In reflecting, in this short position paper, I want to achieve two objectives.

First, to give a personal reflection on the type and trajectory of research around BPMN over the last ten years. This I will do in the section below.

In the subsequent section, I then attempt to identify themes and pathways for research on BPMN that are not fully on the radar screen of the BPMN academic community yet.

As a preamble I do wish to add that what follows are my own thoughts, interpretations and opinions. They are not based on rigorous research but instead based on selective reading, personal experiences and anecdotes.

2   A Brief History of BPMN Research

I have personally been involved with BPMN as a phenomenon of research interest since January 2005. At that time, BPMN was widely discussed if not hyped in the practitioner communities, which in turn raised its profile as a phenomenon of interest to the academic communities in business, information systems and computer science.

This interest, over the past seven years, has increased rather than diminished; but the themes of interest changed. Fig. 1 gives a visual display of my interpretation of the BPMN research themes over the years. Note that this graph is not based on a systematic review of the literature but rather describes a personal sketch of themes and efforts over time.

Two themes dominated the early years: how good is BPMN really? This question was probably academia's answer to the adoption speed and hype that was felt as a buzz in the community. Early work looked to examine actual and perceived capabilities from a variety of angles [17, 22]. Other research at that time examined some of the claims made – BPMN for different stakeholders, and BPMN for organizational redesign versus technology implementation. One key theme was to define the semantics of BPMN formally to identify how the promise of model-driven process execution could be realized [2]. Eventually (of course), an answer was found [8].

Moving forward, BPMN received increased attention and uptake. Capabilities were now well-established and so the community started to look at how these capabilities were implemented and used [24] – but also how they could be extended and improved, leading to input to the ongoing and planned revisions of the standard [12, 23]. One of the noteworthy findings of that time was a study that showed how much of BPMN was used in practice [25] – and what implications could be in terms of standard revision and modeling methodologies and training.

Somewhat linearly following was then research work that begin to these formal, analytical and empirical findings into normative advice – how BPMN should be used. Thus, we saw a set of textbooks and studies emerging around how BPMN was best applied and taught [15, 21]. BPMN also found its way into more general textbooks on Business Process Management [19].

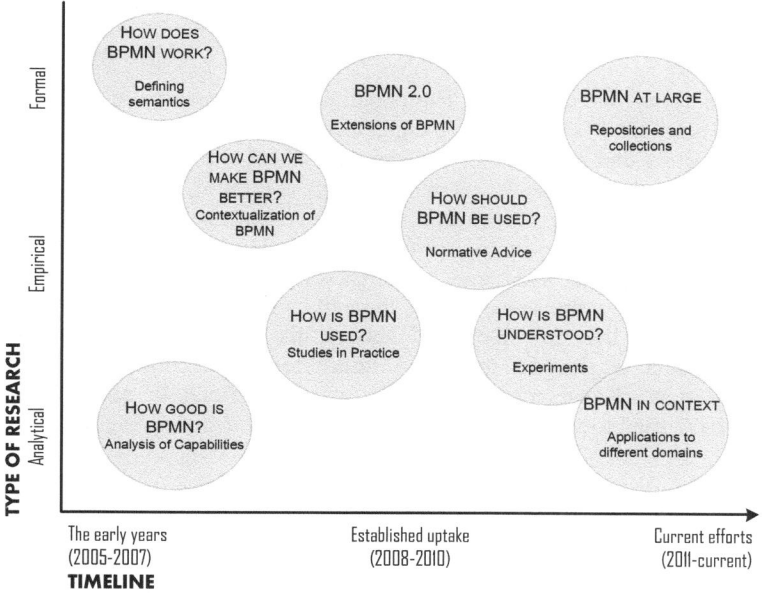

Fig. 1. Selected themes of BPMN research over time

The most recent years saw yet another trend in research around BPMN. BPMN was now firmly established as a technological and organizational approach and thus was used more as a key instantiation of more general approaches– to modeling, say – that were being examined. One example of such research programs is the stream of work that looks at the understandability of process models – perusing BPMN as the models of choice [11]. Other work has started to look at collections of (BPMN) models and the management or technical realization of those [5, 18]. Finally, other work has looked to extend the applications of BPMN to new contexts and domains [6].

I should add that several of the 'theme bubbles' above suggest that a stream of research is complete and finalized. For instance, evaluations and empirical studies of BPMN continue well into the current time [3], as do formalizations and analytical works [13].

With this portfolio existing, alive and well, two questions emerge in my view:

1) Has the research to date been on the right track, i.e., have we focused our efforts on the important challenges?
2) What will be the relevant and important research agenda for the future?

Below, I am offering some thoughts on these two questions.

3   Knowns and Unknowns

There are known knowns; there are things we know we know. We also know there are known unknowns; that is to say we know there are some things we do not know. But there are also unknown unknowns – the ones we don't know we don't know.

Donald Rumsfeld, U. S. Secretary of Defense, Statement to the Press on February 12, 2002.

This section is not about Donald Rumsfeld, his appropriation of a statement that presumably originated from the risk management literature, or the linguistic (let alone political) quarrels that emerged in consequence about the above statement. For the interested reader, the statement was made at a press briefing where Donald Rumsfeld addressed the absence of evidence linking the government of Iraq with the supply of weapons of mass destruction to terrorist groups. He was later criticized for his application of the English language, although some linguists lauded his statement as "impeccable, syntactically, semantically, logically, and rhetorically" [9].

Why do I bring this up? Because I believe his quotation can serve as a useful conceptual frame to examine past and future research around BPMN, and also identify the type of research that can be most beneficial to exploring the ways forward. I have tried to visualize this frame and its application in Fig. 2.

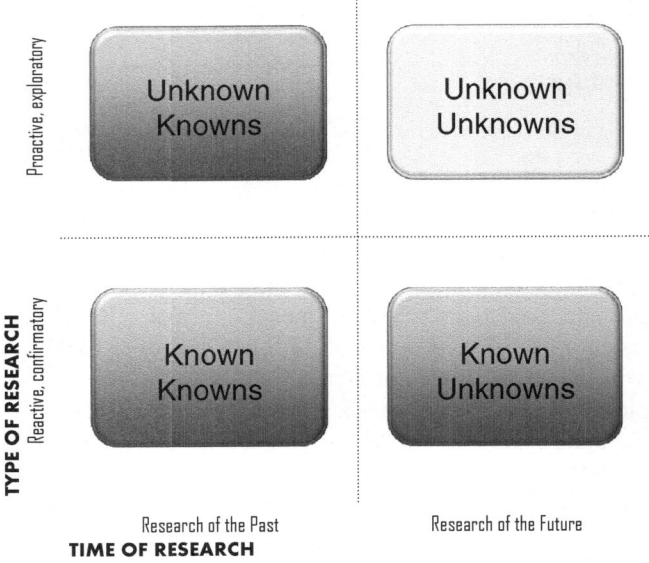

Fig. 2. A taxonomy of BPMN knowledge. Loosely based on Donald Rumsfeld's press statement made as United States Secretary of Defense on 12 February 2002.

The taxonomy in Fig. 2 has two axes. On the x-axis I differentiate *knowns* from *unknowns* with the view to indicate research of the past and the possible themes for the research of the future. The y-axis separates two different types of research – more *reactive* research to *confirm* known knowns and known unknowns – things that we know we know or things that we don't know but at least we know that we don't know them. This type of research – and likely the approach to research – is different from the more *proactive* research required to *explore* the unknowns – things we didn't even know we know, and (perhaps most interestingly), the things we don't know that we don't know. Of course, this last category is the most challenging state to predict – but perhaps also the most interesting to explore?

Perusing this frame, I have tried to instantiate the different categories to guide future research. The outcomes of this effort are described in Table 1.

Table 1. Known and Unknowns

Known Knowns	Known Unknowns
BPMN is used selectively in organizations, and not to its full extent.	The level of errors in BPMN modeling is still high.
BPMN can be mapped to executable semantics.	What is the best way to apply BPMN for process modeling?
There are advantages and disadvantages of BPMN in comparison to other modeling approaches.	The use of BPMN in cultures with different aptitudes for forms, shapes and symbolic expressions.
BPMN is implemented in different ways by process engines.	The process of BPMN development.
	The impact of new technologies for BPMN modeling.
Unknown Knowns	**Unknown Unknowns**
Organizations use BPMN differently for different projects (redesign, implementation, compliance).	How do we use BPMN for different and emerging purposes?
The individual and organizational benefits that flow from BPMN use.	What extensions to the standard will be required in the future?
Defining and implementing workflow systems starting with BPMN models.	Whill BPMN have a place in post-process paradigms?
The use of BPMN by experts and novices.	How will the BPMN community and its impact evolve?

The lists above are incomplete, subjective and debatable by nature. On this we can hopefully agree (to disagree). My point is to be stimulating rather than to be directive or instructing.

Most notably, the framework of known knowns draws our attention to the top right corner of Fig. 2 – the unknown unknowns. Researchers, I believe, are well-suited and even more so, required to conduct work that fits into this quadrant. With their array of methods and knowledge about rigor as well as relevance, researchers should not only explore and confirm knowledge but also proactively design new knowledge and bring implications that guide the development of the whole field for the years ahead. In other words, they are positioned to turn the unknown unknowns into known knowns or at the known unknowns. This move requires boldness and inspirational thinking – to identify areas that are 'way out there' and to identify appropriate ways of executing on such research.

Another key implication from thinking about BPMN research on basis of the known/unknown frame, I believe, is the question of "how do we best go about doing research on/around/with BPMN?". The reactive/proactive distinction I have drawn in Fig. 2 suggests that a variety of streams will be required. Looking back at the research to date (as depicted in Fig. 1), we can identify several camps of studies – formal, analytical, empirical. There are also camps of design and development work. I argue that this diversity of streams is important and required; but also needs to be consolidated, integrated and applied holistically. The different communities of work around BPMN also need to collaborate more tightly: Participants in the academic community, representative from vendor and end user organizations, stakeholders in the standardization bodies and influencers from the teaching community should work together in exploring unknowns and confirming knowns. For us academics, this means that different methodological paradigms appear well-suited that, to date, we have not yet seen fully leveraged or exploited: Mixed method research [4], theory-driven development [1], design research based on participatory action [14] and scholarly work in terms of community engagement [16] are just a number of suggestions to explore and embed in the current portfolio of activities.

Broadening and deepening our perspectives are certainly ways not only to leverage BPMN as an interesting phenomenon for intellectual work, but also to influence positively the different communities impacted by BPMN – scholars, students, vendors, technology innovators, standard bodies and end user organizations alike. From my viewpoint as an academic scholar, I would like to see especially our cohort as leaders in this activity – as boundary spanners that explore, shift and create knowledge of benefits across all parties. Whether this assertion holds, however, is one of Donald Rumsfeld's known unknowns: *"there are some things we do not know."*

References

1. Arazy, O., Kumar, N., Shapira, B.: A Theory-Driven Design Framework for Social Recommender Systems. Journal of the Association for Information Systems 11, 455–490 (2010)
2. Dijkman, R.M., Dumas, M., Ouyang, C.: Semantics and Analysis of Business Process Models in BPMN. Information and Software Technology 50, 1281–1294 (2008)
3. Figl, K., Derntl, M.: The Impact of Perceived Cognitive Effectiveness on Perceived Usefulness of Visual Conceptual Modeling Languages. In: Jeusfeld, M., Delcambre, L., Ling, T.-W. (eds.) ER 2011. LNCS, vol. 6998, pp. 78–91. Springer, Heidelberg (2011)
4. Johnson, R.B., Onwuegbuzie, A.J.: Mixed Methods Research: A Research Paradigm Whose Time Has Come. Educational Researcher 33, 14–26 (2004)
5. La Rosa, M., ter Hofstede, A.H.M., Wohed, P., Reijers, H.A., van der Aalst, W.M.P.: Managing Process Model Complexity via Concrete Syntax Modifications. IEEE Transactions on Industrial Informatics 7, 255–265 (2011)
6. Müller, R., Rogge-Solti, A.: BPMN for Healthcare Processes. In: 3rd Central-European Workshop on Services and their Composition, pp. 65–72. CEUR-WS.org, Karlsruhe (2011)
7. OMG: Business Process Model and Notation (BPMN) - Version 2.0. Object Management Group (2010), http://www.omg.org/spec/BPMN/2.0

8. Ouyang, C., van der Aalst, W.M.P., Dumas, M., ter Hofstede, A.H.M., Mendling, J.: From Business Process Models to Process-Oriented Software Systems. ACM Transactions on Software Engineering Methodology 19, 2–37 (2009)
9. Pullum, G.K.: Language Log: No Foot in Mouth (2003), http://itre.cis.upenn.edu/~myl/languagelog/archives/000182.html
10. Recker, J.: Opportunities and Constraints: The Current Struggle with BPMN. Business Process Management Journal 16, 181–201 (2010)
11. Recker, J., Dreiling, A.: The Effects of Content Presentation Format and User Characteristics on Novice Developers' Understanding of Process Models. Communications of the Association for Information Systems 28, 65–84 (2011)
12. Rodríguez, A., Fernández-Medina, E., Piattini, M.: A BPMN Extension for the Modeling of Security Requirements in Business Processes. IEICE Transactions on Information Systems 90, 745–752 (2007)
13. Sanchez-Gonzalez, L., Ruiz, F., Garcia, F., Cardoso, J.: Towards Thresholds of Control Flow Complexity Measures for BPMN Models. In: 26th ACM Symposium on Applied Computing, pp. 1445–1450. ACM, TaiChung (2011)
14. Sein, M.K., Henfridsson, O., Purao, S., Rossi, M., Lindgren, R.: Action Design Research. MIS Quarterly 35 (2011)
15. Silver, B.: BPMN Method and Style: A Levels-based Methodology for BPM Process Modeling and Improvement Using BPMN 2.0. Cody-Cassidy Press (2009)
16. Van de Ven, A.H.: Engaged Scholarship: A Guide for Organizational and Social Research. Oxford University Press, New York (2007)
17. Wahl, T., Sindre, G.: An Analytical Evaluation of BPMN Using a Semiotic Quality Framework. In: Castro, J., Teniente, E. (eds.) CAiSE 2005 Workshops, vol. 1, pp. 533–544. FEUP, Porto (2005)
18. Weber, B., Reichert, M., Mendling, J., Reijers, H.A.: Refactoring Large Process Model Repositories. Computers in Industry 62, 467–486 (2011)
19. Weske, M.: Business Process Management: Concepts, Languages, Architectures. Springer, Berlin (2007)
20. White, S.A.: Introduction to BPMN. BPMI.org (2006), http://www.bpmn.org/Documents/Introduction%20to%20BPMN.pdf
21. White, S.A., Miers, D.: BPMN Modeling and Reference Guide. Future Strategies, Lighthouse Point (2008)
22. Wohed, P., van der Aalst, W.M.P., Dumas, M., ter Hofstede, A.H.M., Russell, N.: On the Suitability of BPMN for Business Process Modelling. In: Dustdar, S., Fiadeiro, J.L., Sheth, A.P. (eds.) BPM 2006. LNCS, vol. 4102, pp. 161–176. Springer, Heidelberg (2006)
23. Wolter, C., Schaad, A.: Modeling of Task-based Authorization Constraints in BPMN. In: Alonso, G., Dadam, P., Rosemann, M. (eds.) BPM 2007. LNCS, vol. 4714, pp. 64–79. Springer, Heidelberg (2007)
24. zur Muehlen, M., Ho, D.T.-Y.: Service Process Innovation: A Case Study of BPMN in Practice. In: Sprague Jr., R.H. (ed.) Proceedings of the 41th Annual Hawaii International Conference on System Sciences, p. 372. IEEE, Waikoloa (2008)
25. zur Muehlen, M., Recker, J.: How Much Language Is Enough? Theoretical and Practical Use of the Business Process Modeling Notation. In: Bellahsène, Z., Léonard, M. (eds.) CAiSE 2008. LNCS, vol. 5074, pp. 465–479. Springer, Heidelberg (2008)

A Platform for Research
on Process Model Collections

Rami-Habib Eid-Sabbagh, Matthias Kunze,
Andreas Meyer, and Mathias Weske

Hasso Plattner Institute at the University of Potsdam
Prof.-Dr.-Helmert-Strasse 2-3, 14482 Potsdam, Germany
{rami.eid-sabbagh,matthias.kunze,
andreas.meyer,mathias.weske}@hpi.uni-potsdam.de

Abstract. Business process management has received considerable attention and many companies achieved a high maturity level and hence, generated collections of process models that form a knowledge asset essential to their operations. These collections bear opportunities for innovation: Empirical research establishes methods and techniques to support and improve business process management; yet, these need to be validated with regards to process models from industry. However, due to their heterogeneity, extracting and analyzing process models from process model collections is a tedious task and time consuming.

To facilitate access to process model collections, this paper presents an extensible platform for their analysis that shall support researchers and foster collaboration and reuse. This platform provides importing functionality for a set of process collections recognized in research and functionality to easily explore, transform, and extract information from process model collections. In a small showcase, we illustrate the application of the platform towards clustering a collection of process models.

1 Introduction

Business process management (BPM) is central to modern organizations, since it provides established methods and techniques to document, control, and evaluate the operations carried out. The essential assets of BPM are process models, as they capture how work is performed and how goals are achieved. Further, they sustain, among others, communication, automation, certification, and performance evaluation of business processes. Consequently, these organizations collect hundreds or thousands of process models [1,2].

Within the last decade, BPM became a mature field with regards to both practical application and scientific research. Many approaches and algorithms towards empirical research in the field of BPM have been proposed along with several process model collections that have been made available by companies [3,4] and public bodies [5,6]. While these collections have been subject to validate research results, it is difficult for researchers to apply their work to different collections—a main obstacle that hinders empirical evaluation. This is due to

J. Mendling and M. Weidlich (Eds.): BPMN 2012, LNBIP 125, pp. 8–22, 2012.
© Springer-Verlag Berlin Heidelberg 2012

the heterogeneity of the structure of these collections, their format, and modeling language, which yield a considerable overhead to explore, transform, and extract relevant information.

To reduce work and foster empirical research in the context of process model collections, we developed an analysis platform that aims at allowing researchers to focus on their actual research questions, rather than spending time and effort preprocessing data from heterogeneous sources. In this paper, we discuss the scientific context and present the platform's design and implementation.

The platform provides uniform access to a large variety of process model collections, by means of import functionality that allows capturing any kind of information related to a process model. Utilities facilitate iterating over process models, filtering them, segmentation and extraction of information, as well as transformation into a generic representation that covers common concepts of process modeling languages. As the platform is an open source project[1], maintained by a scientific community, it is accessible to everyone and open for extensions.

Please note that the proposed platform is not a process repository to actively manage process model collections, but an analysis platform, which provides the fundamental infrastructure (including an uniform access to different process model collections) for analysis. Currently a small set of analysis modules is provided, for instance, process metrics calculation as well as process collection clustering capabilities. The set of the platform's analysis modules will be extended by researchers sharing their developed analysis modules for the platform. With the increase of shared analysis modules and process collection importers, implementation work for the individual researcher dealing with a particular interest decreases as existing modules can easily be reused.

The remainder of this paper is structured as follows. Section 2 establishes the context of empirical research on process model collections and discusses related work, before the design and implementation of the platform are presented in detail in Section 3. Based on that, Section 4 briefly illustrates a showcase towards analyzing and clustering process models from a large collection with the help of utilities the platform provides. Section 5 concludes the paper and gives an outlook on current and future work.

2   An Ecosystem of Empirical Research

The term ecosystem refers to the interaction of individuals in a particular environment. In this section, we present the ecosystem that embeds practice and empirical research with regards to process model collections, outlined in Fig. 1. Typically, there is only little direct interaction between the individuals, i.e., experts, from practice and research. However, as mentioned above, research largely depends on the information constituted during the practical application of BPM—in our case collections of process models.

Experts from the *practice* area are domain and process modeling experts that create, reuse, and manage process models in large collections [1,2], or business

---

[1] http://bpmai.org/BPMAcademicInitiative/BpmTools

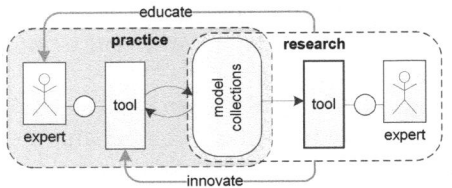

Fig. 1. The ecosystem of process model collections

analysts that examine process performance. For every particular task, these experts require specific tools. For instance, domain experts need a process modeling tool that facilitates conceptual elicitation of a process and supports the modeler while abstracting from technical details. In Fig. 1, this is represented by the human actor that communicates with a tool to access and manipulate process collections.

Researchers suggest improvements to the practical side through innovation and education. Within the last few years, growing interest has influenced research work towards large process model collections, addressing methods, e.g., [7,8], and tool support, e.g., [9,10], in the field of BPM. Achievement of such results demands access to methods applied and actual process models created by practitioners, depicted on the right side of Fig. 1.

In the center of practice and research stand *process model collections*. To date, there are only few such collections available for research. One of the most commonly used collections is the SAP reference model [3], published in 1997, that comprises 604 EPC process diagrams that describe the SAP R/3 system. More recently, IBM released[2] a collection of 735 business process models from the Websphere process modeler available as BPMN process models, which cover various industrial domains, e.g., finance and telecommunications, [4]. In the course of the BPM academic initiative[3] (BPM AI), a set of 1903 process models as of time of writing this paper (June 2012) created by students in various process modeling languages has been made available to the research community [11].

Research of large process model collections bears significant potential for BPM practitioners [12], which we briefly present in the context of above ecosystem.

Educate. From quantitative analysis of many process models, one can draw conclusions that influence the way, experts are using process models. Related work [13,11] showed that the majority of models uses only few distinct modeling constructs and is of low complexity, which can guide in reducing the spectrum of modeling languages and hence, ease and accelerate process model design. Guidelines towards process modeling [14,15] and labeling [16] suggest to reduce error proneness and increase model understanding.

Large collections allow identification of common modeling styles or commonly reused process models, which is a prerequisite to the development of reference models for business processes [17]. For example, the MIT process handbook [18]

---

[2] http://www.zurich.ibm.com/csc/bit/downloads.html
[3] http://bpmai.org

provides an ontology of terms and their relations in the context of business processes.

Innovate. On the other hand, a lot of research addresses the development of functionality used for effective process model management. At the basis of these endeavors stand process model repositories that provide structured access to stored process models and support process specific tasks, e.g., life cycle management, [6]. One particular aspect of managing process model collections is process model search [19], because effective reuse requires meaningful search capabilities. With regard to this, [20] compares the effectiveness of various measures towards process model similarity, whereas [21] showed that similarity metrics can be applied to perform search very efficiently. Further approaches focus on disseminating models into fragments [22] or potential execution paths [23] and search for their occurrence in stored models.

A relatively young topic of process model management is process architectures [12]. Process collections are typically only structured along one perspective, e.g., a functional view, whereas different users or tasks require a variety of views on a process model collection. The term process architecture refers to the relations between processes in a collection and guidelines to organize them [24]. Recent and ongoing research aims at providing perspectives on process architectures that serve particular purposes, respectively; also addressing automatic techniques to derive such structures [25].

At the center of this paper stand software tools that assist the researcher in disseminating and analyzing large process model collections, depicted by the *tool* actor in the right part of Fig. 1. Typically, these tools are implemented specifically for a particular and narrow task, and cannot easily be reused for other process model collections. This is due to process model collections leveraging different structures and formats of process models. Often, the same functionality is implemented repeatedly by several researchers. Therefore, we propose a platform that provides uniform access to different process model collections, along with import and parsing functionality for a set of existing process model collections. Researchers are encouraged to implement their analysis tools on top of this platform, repeat their experiments easily with other collections, and share their methods and insights with the research community.

Although the presented platform provides utilities to access process model collections, it is not intended for collection management similar to process model repositories, cf. [6]. In particular, repositories address the support of process model life cycles, i.e., creation, update, relationships among models, etc, whereas we assume process model collections to be a rather static basis to conduct analyses upon. For example, Apromore [26] provides functionality to evaluate, filter, design, and present process models and manage large collections thereof, which in part resembles analysis conducted with our platform. However, the main focus of our platform is not to provide analysis algorithms, but to support researchers in applying their empirical studies and to facilitating reuse of research and analysis techniques across different process model collections. The following section will explain, how this is achieved on a technical basis.

3   Platform Design

In this section, we present the design of our platform, beginning with the requirements for research on process model collections. Based on them, the architecture and core features are illustrated in detail below.

3.1   Requirements

To provide a platform for research on process model collections bears several challenges and particular requirements to the software architecture of the system. The requirements for our platform were derived from the ecosystem illustrated in Section 2, i.e., various process collections that differ in structure, process model representation, and content on the one hand, as well as current research questions on the other hand.

Management of Heterogeneous Process Model Collections. Process collections have been created with different intentions depending on the organizational domain and on the modelers' level of BPM expertise. As a consequence, process collections differ, among others, in process modeling languages, e.g., BPMN, EPC, and Petri nets, file format, and accompanying metadata. For instance, all models from the SAP reference model are stored in one large EPML file; each model from the BPM AI, in contrast, is represented by a JSON file that captures the model graph and an SVG file that contains a visual representation of the model. In the national process library [5], many process models have no graphical representation at all, but they are classified by a set of structured metadata and described in prose. During the design of this platform, we also envisioned processes to be represented solely by execution logs rather than any model description.

Hence, a universal process analysis platform, which does not restrict research questions asked, should not prescribe any meta-model, but capture all kinds of process representations. As one process model can appear in different forms, e.g., as in the case of BPM AI with JSON and SVG representations, the platform shall be able to store several representations of one process model and always keep the original sources.

Versioning engineered artifacts is good practice in order to track changes and revert mistakes; this also holds true for process models and is supported by several BPM tools. Hence, it is compulsory for our platform to preserve version information, too.

Filtering, Segmentation, Storage, and Modular Analysis. Above characterization of the heterogeneity of process model collections requires a modular approach to discover relevant aspects of process models in a large collection for analysis. Therefore, it is desirable to filter the set of process models of one or several process collections, transform them into a uniform representation, and extract particular features (segmentation) for consideration. This allows, for instance, obtaining only activity labels of process models from the financial domain before applying custom analysis features.

Analysis of process models and collections should be modular in order to facilitate reuse of certain functionality across different experiments and allow researchers to benefit from each other more effectively. By this, more complex analyses can be constructed by assembling several atomic analysis modules and execute them over a filtered and segmented set of process model data.

Researchers, who work on large data sets and computation intensive tasks, may run experiments over several hours or days. As we expect an increasing amount of data that analyses shall be applied to, the platform also needs capabilities to store analysis results relating them to particular process models and providing means to efficiently obtain this data again.

Support Reuse and Repeatability. Research typically builds on former research results, refers to it, extends it, and improves it. In a similar way, the platform shall allow easy (re-)use of analysis functionality and help researchers to cooperate with each other. Modular analysis modules are one step into this direction.

Methods or innovations derived from analyzing one process model collection are only valid for the context of that collection, i.e., they may not be valid in other cases. Hence, external validation beyond a single collection requires application of the same procedures to other collections. Here, a uniform representation of graph-based process models provides a solid basis for analysis. This enhances the platform to use one format to iterate over different process languages on the one hand. On the other hand, analysis modules can be re-used in an easy way and do not need to be re-implemented for different process representations. Thus, a structure that is flexible enough to map different process modeling notations but not too generic to reduce them to meaningless blocks is required.

Nevertheless, the original representation of process models must be retained to analyze aspects specific to the representation of a process model or if a mapping cannot be provided, e.g., in case the process is only described textually.

3.2   Architecture

According to above requirements, we identify three main functional areas in the conceptual architecture of our platform, illustrated in Fig. 2: Import, analysis, and index management. After a brief introduction of these features in the context of the architecture, we will highlight each area in detail hereafter.

The *import* module extracts distinct process models from the original collection and imports them into the platform, mapping them to a flexible data schema to capture all information provided. *Filter management* contains compact units of functionality to iterate over imported process model collections, filter, segment, and transform process models and supply the extracted information into the *analysis* modules provided by the respective researcher. Finally, *index management* allows storing analysis results related to process models in a key-value store and obtaining stored data by means of *querying* that data.

In addition to the core functionality, the platform relies on a *process model repository* to store the process models and analysis results. The actual

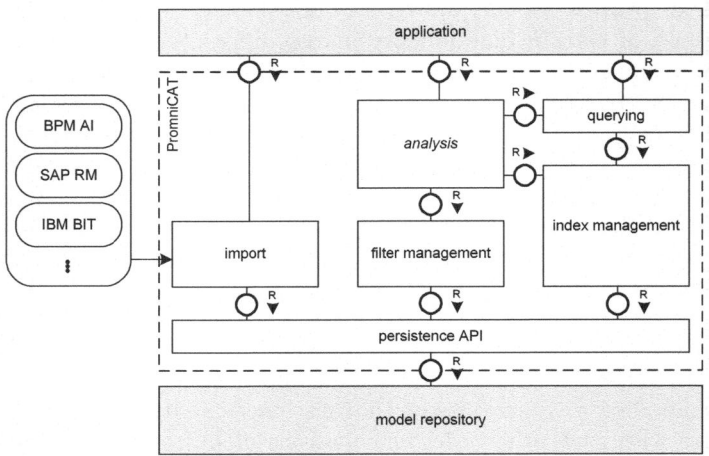

Fig. 2. System architecture of the platform

repository interface is wrapped by the *persistence API* that maps interfaces offered by the repository to interfaces required by the platform's modules. In the present case, we resorted to a graph-document database management system, that is, OrientDB[4]. Theoretically, also existing process model repositories, e.g., Apromore [26], could be leveraged instead.

As mentioned before, we do not provide an analysis tool, but a platform to build analysis functionality on top. Hence, complete *applications* could be built leveraging the capabilities of our platform. Such application may directly interact with the importer, analysis, and index management modules.

Data Schema and Import. The data schema provided by the persistence API is presented in Fig. 3. Here, a *model* stores all information that is related with one logical process model from the collection. Each model can have several *revisions*, based on our observation that each process model can be changed over time and that these changes may be relevant to research. Each revision, in turn, can have any number of *representations*.

For each process model, its *title*, *origin*, and an identifier (*importedId*) are stored. The origin indicates the process model collection from where a model has been imported. The *importedId* is used to correlate imported models with models from the collection.

Revisions store different milestones of the evolution of a model. If the process modeling tool, used to create models in the collection, allows storing several versions of the model, these should not be treated as distinct models, but correlated with the same model. If models are not versioned in the collection, a model will be linked to exactly one revision in our data model. Each revision is identified by a *revision number*, its *author*, and a flag (*latestRevision*) that identifies the most recent revision.

---

[4] http:\www.orienttechnologies.com

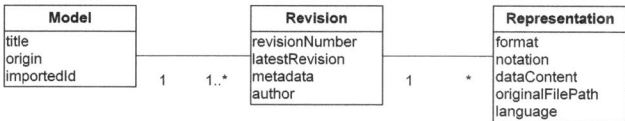

Fig. 3. Process model data schema

As introduced in Section 3.1, we do not restrict the type of data to be analyzed. Hence, our platform allows importing almost any kind of process descriptions independently from their format, structure, properties, and the capability to be mapped to a generic process representation. This is achieved by relating a process model with one or many representations, i.e., process descriptions in their original format, see Fig. 3. More precisely, every revision can link different representations and classify them according to their file *format*, the modeling *notation* used, and, if provided, the natural *language* that is used in the model inscriptions. Since we do not restrict the data to be stored, the original process representation imported from the collection is stored as byte array in our model repository in *dataContent*.

In some cases, models may have no representation or a presentation is complemented by structured data, e.g., in the national process library [5], see Section 3.1. Hence, a revision can also hold structured data in *metadata*.

As various process collections are structured differently, a separate import module is required for every kind of process collection to be imported. This importer needs to identify several revisions of the same model, group them together, and discover different representations, e.g., a bitmap representation for visualization and a graph representation to perform analyses. For each representation, the module extracts its content, analyzes it towards certain aspects, i.e., its file format, its natural language, the applied process modeling language (notation), and relates it with a revision.

Currently, the platform provides importers for the process collections presented in Section 2, i.e., an EPML importer (for the SAP reference model), an importer for BPMN models in the standard XML representation (for IBM BIT models), and an importer for models from the Signavio process modeler that has been used to create models of the BPM AI. Additionally, we also allow importing from the national process library, which mainly contains process descriptions in natural language text, sometimes extended by raster image representations.

Importers also support update functionality to synchronize an already imported process model collection with a more recent version of this collection. This is realized by correlating collections through the *origin* attribute of a model and identifying existing models by their *importedId*, which denotes a unique identifier within a process collection. An update would then only add new models, new revisions, and additional representations, rather than changing stored representations to avoid invalidating existing analysis results. However, it is also possible to import a previously imported collection again by giving it a new *origin*.

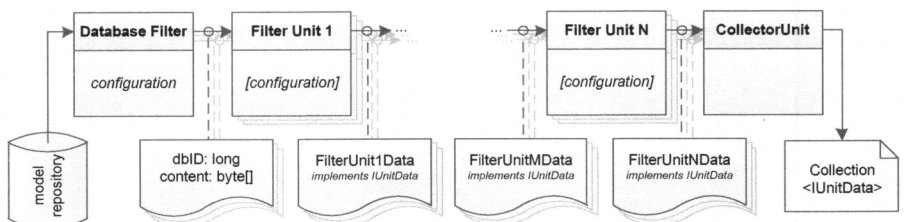

Fig. 4. Example filter chain that extracts labels from a BPMN process models

Model Analysis. The central purpose of our platform is to provide a fundament for model analysis, realized by the interaction of *filter management* and *analysis*, illustrated in Fig. 2. Technically, the analysis module is rather a collection of several units that offer particular algorithms to examine models, sets of models, or aspects thereof. These analysis units are to be developed by researchers with regard to their research question and requirements. Nevertheless, we already provide some analysis units, e.g., a subset of the metrics that measure process complexity presented in [14].

To release researchers from the burden to choose models from a collection, extract relevant information, and transform it into a format applicable for the respective analysis units, the platform provides so-called filter chains, following the pipes and filter integration pattern [27]. Here, filters serve as utilities to load and handle process models from the repository and can be categorized by their function to filter, transform, or extract information.

A schematic example for filter chains is given in Fig. 4; a concrete example is given in Section 4. Access to the repository is always provided by a *DatabaseFilter* that allows setting filter criteria on the data model presented in Fig. 3. For each model, the database filter provides its database id and the loaded model representation. Attached to the *DatabaseFilter* follow *Filter Units* that accept the output data from the previous filter as their input, select, transform, or extract content, and provide their result to the subsequent filter. Input and output data of a filter represent information extracted from one model, and must implement the interface *IUnitData*. Only filters that match in their input and output data type can be connected. For each model, this chain is separately executed, and several chains can be executed in parallel to increase processing performance on multi-CPU computers. Finally, the *collector* consolidates the results of every chain and provides them to the analysis module.

The platform already provides a set of filter units readily usable by researchers, for instance, the aforementioned database filter. Further units provide functionality to segment process models into particular concepts, e.g., extract all activities of a process model.

As we encourage reuse of implemented functionality within the platform, we envision analysis modules to be implemented as generic as possible. By that, analysis can easily be applied to different process model collections to repeat previous experiments and compare results. That is, whenever possible, analysis modules should be built on a generic process representation. We see that many process modeling languages, e.g., BPMN, EPC, and UML Activity Diagrams, share common concepts, e.g., activities, events, gateways, sequence flow, data objects, and resources. Hence, research that addresses these concepts can benefit from a given generic format and parsers that transform stored processes into this representation. This transformation is performed by a utility unit as well.

Therefore, our platform provides a filter unit that parses process models to a representation in the *jbpt* format. Jbpt is a Java-library[5] that leverages graph structures to support a canonical process representation providing descendants for each supported modeling language. Algorithms that work on the canonical representation can also be applied to descendants. Besides, this library offers a comprehensive set of techniques to transform, verify, and analyze process models, e.g., provide transformation of process models to Petri nets, soundness checking, net unfoldings, or workflow graph decomposition into process structure trees.

Again, for every type of representation that shall be imported, a corresponding parser is required that transforms the model into the jbpt graph format. Currently, transformations for EPC and BPMN process models from the SAP reference model, IBM BIT, and BPM AI model collections are available. If a representation cannot be transformed to jbpt, e.g., in case of textual or bitmap process descriptions or in absence of a representation, filter units can extract relevant aspects and analysis units need to account directly for proprietary formats.

Indexes. As mentioned before, we perceived the need to store analysis results within the platform by simple means. On the one hand, this frees researchers from the need to maintain their own persistence mechanism to store results; on the other hand, it allows referring to stored process models while maintaining consistency of linked models.

As the platform addresses analysis of process model collections, we envision analysis results also to be collections of data, likely categorized along certain dimensions. Therefore, we opted for a key-value store that allows relating any kind of Java data to a key that is either a string or a number. Due to the character of this $m{:}n$ mapping, where one key may relate several data entities and one data entity may be identified via several keys, we refer to the storage as an index. The querying module provides means to select a subset of stored data by querying over keys. Querying also offers to compute the intersection between distinct indexes.

However, this mechanism is not intended to support fast process model indexing and search but rather provide a simple interface to store and access intermediary and result data obtained during the analysis of process model collections. Hence, it is up to the model repository to efficiently implement access to this storage.

---

[5] http://code.google.com/p/jbpt/

4   Show Case

In this section, we present an example with technical details on the use of our platform. [25] presents an approach to create navigation structures for an unstructured process collection by using hierarchical clustering. Utilizing this approach, we focus on the support the platform provides to access and extract desired elements from process collections rather than on the clustering algorithm itself, for which the elements are extracted.

Structuring a process collection using clustering can be performed in three main steps: (1) preprocessing process model data, (2) clustering using a similarity function, and (3) labeling of clusters. The preprocessing step of this approach demonstrates the ease of using the platform and will be our main interest.

In this example, we like to extract labels of BPMN process model activities from the BPM AI collection that shall be used for clustering. To perform these steps, we need to set up a unit chain with the *UnitChainBuilder*. Fig.5 illustrates the unit chain to extract the activity labels from each process model and to calculate the corresponding feature vector depicting the input and output of each unit chain element. All utility units implement the *IUnitData* interface. Not all of the filter utility units can be put in direct sequence, because each consumes specific input and produces specific output data. The platform ensures compatibility of input and output for directly succeeding units.

First, we need to create and configure the database filter. This is done by creating a new *DbFilterConfig* and adding origin, format, and notation to it. Origin refers to the process model collection, format refers to the desired representation of the models in the database, and notation requests the process

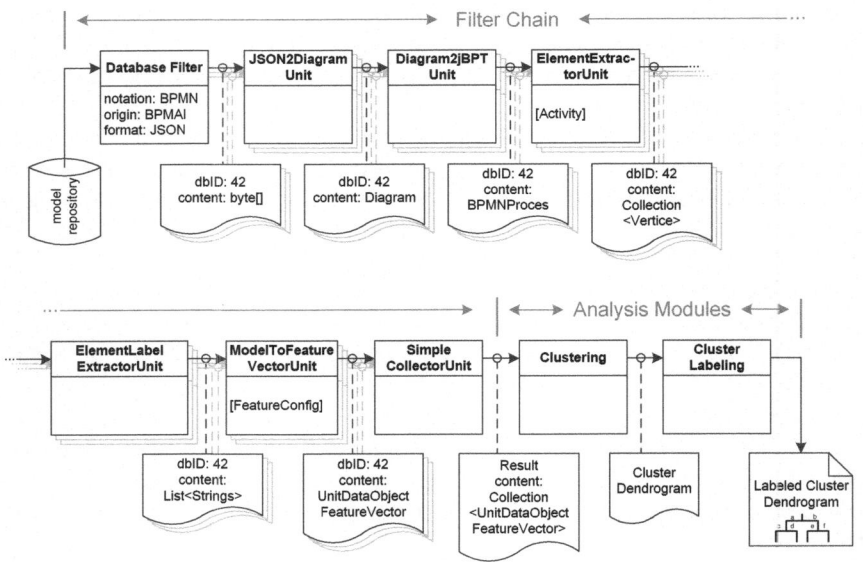

Fig. 5. Show case unit chain and analysis modules

modeling language. As we are interested in BPMN process models from the BPM AI collection, we set *BPMAI* as origin value, *BPMAI_JSON* as format value, and *BPMN* as notation value accordingly. If the *DbFilterConfig* is not set, all process models from all collections will be extracted.

Then, the database filter is added to the *UnitChainBuilder*. With the above configuration of the database filter, 1449 BPMN models are extracted from a total of 1903 models in the BPM AI collection.

In the next step, the process models must be mapped into the generic jbpt format to enable the use of existing activity and label extraction utility filters. Therefore, we use the method *BpmaiJsonToJbpt*. This method executes the *BpmaiJsonToDiagramUnit* and the *DiagramToJbptUnit* in sequence to transform the JSON representation of the process model into a jbpt process model representation. The *BpmaiJsonToDiagramUnit* input type is *String* and provides a *Diagram* as output. The *DiagramToJbptUnit* reads this output and converts the diagram into a jbpt process model, provided at its output interface.

Process models in the jbpt format can now be queried for specific elements of interest. In our case, we want to extract all activities from each process model. This is done by adding an *ElementExtractorUnit* to the chain. Its input is a jbpt process model, which we created in the last step, and its output is a collection of vertices, because BPMN activities inherit from a vertex type in jbpt. At this point, we have extracted a set of 11600 activities from the chosen BPMN models.

To extract the labels from the activities, we add the *ElementLabelExtractorUnit* that performs this task. As a result we receive a list of labels serving as input to set up a feature vector for each process model, i.e., prepare the labels for clustering. So far, we only used existing functionalities of the platform. But for the current step, a *ModelToFeatureVectorUnit* needed to be implemented implemented. This utility unit creates a feature vector of each process model according to its configuration, so that also other values than labels can be prepared for clustering.

Finally, a *SimpleCollectorUnit*, provided by the platform, is added to the unit chain to collect and consolidate the result of the utilized unit chain. In each step of the unit chain, the reference to the original process model is preserved. Listing 1 shows the code required to set up above filter chain and demonstrates that only little effort is required to access the desired aspects of process model collections for analysis.

Analysis is performed by a custom clustering analysis module. This module extends the hierarchical cluster algorithm from the WEKA [28] machine learning tool to deal with strings. For this, the labels are converted into numerical values and then fed to the hierarchical cluster algorithm. The output is a hierarchical structure tree, a so-called dendrogram.

Having a hierarchical tree structure in form of a dendrogram, a navigation structure for a process collection is given. However, without labels of the clusters, this is of limited use. Labeling the clusters provides context to the user while browsing through a process collection. The cluster labeling is performed by analyzing all labels of each cluster and identifying the words with the highest frequency, of which the two most frequent ones are chosen as labels. For illustrative purposes, the technique described here is rather simple. More sophisticated algorithms can be applied by replacing the analysis modules with more complex ones.

```
1   IUnitChainBuilder chainBuilder = new UnitChainBuilder(..., ...,
        UnitDataLabelFilter.class);
2
3       DbFilterConfig dbFilter = new DbFilterConfig();
4       dbFilter.addOrigin(Constants.ORIGINS.BPMAI);
5       dbFilter.addFormat(Constants.FORMATS.BPMAI_JSON);
6       dbFilter.setLatestRevisionsOnly(true);
7       chainBuilder.addDbFilterConfig(dbFilter);
8
9       //transform to jbpt and extract activity labels
10      chainBuilder.createBpmaiJsonToJbptUnit();
11      chainBuilder.createElementExtractorUnit(Activity.class);
12      chainBuilder.createElementLabelExtractorUnit();
13      chainBuilder.createModelToFeatureVectorUnit(featureConfig);
14
15      //collect results
16      chainBuilder.createSimpleCollectorUnit();
17
18      // run chain
19      Collection<IUnitDataLabelFilter<Object>> result = chainBuilder.
            getChain().execute();
```

Listing 1. Code excerpt for filter chain to extract activity labels

5 Conclusion

With the availability of an increased number of large process model collections of the private and public sector, we have observed a shift of focus of the research community to the examination of process model collections. Nowadays, few large collections like the SAP reference model, the IBM BIT models, or the BPM AI collection form a common set for research.

Analyzing these large data sets bears potential for empirical validation, innovation, and education. Within the ecosystem of interaction between BPMN experts from practice and research, problems are posed by BPM practitioners and improvements suggested by researchers in an innovation cycle. The results of these analyses are reflected in new functionalities, hence better process model tools, and education of people in regard to business process management with its various fields of interest.

To support the exchange of practice and research and especially to support the scientific community in their research, we proposed an analysis platform that allows the import of various process model collections into one system in this paper. In this regard, we presented the conceptual design of the platform as well as its concrete implementation. The platform allows researchers to focus on their research by offering importers for a common set of process model collections as well as the openness to develop importers for process model collections becoming newly available. By providing filters to easily filter, extract, and transform process models from large collections, the platform helps researchers to focus on the elaboration of their analysis instead of spending valuable time to implement custom logic for these tasks.

The concept of utility chains that consists of a range of filters and connected analysis modules shall foster re-use and repeatability of implemented functionality within the platform. A mapping to a generic process representation facilitates researchers to run their experiments over several process collections. Therefore, analysis can easily be applied to different process model collections to repeat previous experiments and compare results.

In a brief example of clustering process models in a process collection, we demonstrated the use of the platform and its utility units, with a focus on the filter functionalities provided by the platform.

Currently, we work on implementing a process-centric survey platform to connect to the platform's application interfaces. The survey platform shall be used to validate implemented analyses in user studies or to get feedback on process models from users. The research infrastructure presented in this paper allows a wide range of uses. Various extensions of the platform are envisioned and may be examined in future work.

Acknowledgments. The authors thank the students Cindy Fähnrich, Tobias Hoppe, and Andrina Mascher for their contribution and commitment to the design and implementation of this platform.

References

1. Rosemann, M.: Potential pitfalls of process modeling: part B. Business Process Management Journal 12(3), 377–384 (2006)
2. Akkiraju, R., Ivan, A.: Discovering Business Process Similarities: An Empirical Study with SAP Best Practice Business Processes. In: Maglio, P.P., Weske, M., Yang, J., Fantinato, M. (eds.) ICSOC 2010. LNCS, vol. 6470, pp. 515–526. Springer, Heidelberg (2010)
3. Curran, T., Keller, G., Ladd, A.: SAP R/3 Business Blueprint: Understanding the Business Process Reference Model. Prentice-Hall, Inc., Upper Saddle River (1997)
4. Fahland, D., Favre, C., Jobstmann, B., Koehler, J., Lohmann, N., Völzer, H., Wolf, K.: Instantaneous Soundness Checking of Industrial Business Process Models. In: Dayal, U., Eder, J., Koehler, J., Reijers, H.A. (eds.) BPM 2009. LNCS, vol. 5701, pp. 278–293. Springer, Heidelberg (2009)
5. Eid-Sabbagh, R.H., Kunze, M., Weske, M.: An Open Process Model Library. In: Proceedings of the 1st International Workshop on Process Model Collections, PMC 2011 (2011)
6. Yan, Z., Dijkman, R.M., Grefen, P.W.P.J.: Business Process Model Repositories - Framework and Survey. Information & Software Technology 54(4), 380–395 (2012)
7. van der Aalst, W.M.P.: Process Mining - Discovery, Conformance and Enhancement of Business Processes. Springer (2011)
8. Weske, M.: Business Process Management: Concepts, Languages, Architectures, 2nd edn. Springer (2012)
9. Decker, G., Overdick, H., Weske, M.: Oryx – An Open Modeling Platform for the BPM Community. In: Dumas, M., Reichert, M., Shan, M.-C. (eds.) BPM 2008. LNCS, vol. 5240, pp. 382–385. Springer, Heidelberg (2008)

10. ter Hofstede, A.H.M., van der Aalst, W.M.P., Adams, M., Russell, N.: Modern Business Process Automation - YAWL and its Support Environment. Springer (2010)
11. Kunze, M., Luebbe, A., Weidlich, M., Weske, M.: Towards Understanding Process Modeling – The Case of the BPM Academic Initiative. In: Dijkman, R., Hofstetter, J., Koehler, J. (eds.) BPMN 2011. LNBIP, vol. 95, pp. 44–58. Springer, Heidelberg (2011)
12. Dijkman, R.M., Rosa, M.L., Reijers, H.A.: Managing large collections of business process models - current techniques and challenges. Computers in Industry 63(2), 91–97 (2012)
13. Muehlen, M.Z., Recker, J.: How Much Language Is Enough? Theoretical and Practical Use of the Business Process Modeling Notation. In: Bellahsène, Z., Léonard, M. (eds.) CAiSE 2008. LNCS, vol. 5074, pp. 465–479. Springer, Heidelberg (2008)
14. Mendling, J.: Metrics for Process Models: Empirical Foundations of Verification, Error Prediction, and Guidelines for Correctness. LNBIP, vol. 6. Springer, Heidelberg (2008)
15. Mendling, J., Reijers, H.A., van der Aalst, W.M.P.: Seven process modeling guidelines (7pmg). Inf. Softw. Technol. 52(2), 127–136 (2010)
16. Mendling, J., Reijers, H.A., Recker, J.: Activity Labeling in Process Modeling: Empirical Insights and Recommendations. Inf. Syst. 35(4), 467–482 (2010)
17. Fettke, P., Loos, P., Zwicker, J.: Business Process Reference Models: Survey and Classification. In: Bussler, C.J., Haller, A. (eds.) BPM 2005. LNCS, vol. 3812, pp. 469–483. Springer, Heidelberg (2006)
18. Malone, T., Crowston, K., Herman, G.: Organizing Business Knowledge: The Mit Process Handbook. Process Handbook. MIT Press (2003)
19. Dumas, M., García-Bañuelos, L., Dijkman, R.M.: Similarity Search of Business Process Models. IEEE Data Eng. Bull. 32(3), 23–28 (2009)
20. Dijkman, R., Dumas, M., van Dongen, B., Käärik, R., Mendling, J.: Similarity of business process models: Metrics and evaluation. Information Systems 36(2), 498–516 (2011); Special Issue: Semantic Integration of Data, Multimedia, and Services
21. Kunze, M., Weidlich, M., Weske, M.: Behavioral Similarity – A Proper Metric. In: Rinderle-Ma, S., Toumani, F., Wolf, K. (eds.) BPM 2011. LNCS, vol. 6896, pp. 166–181. Springer, Heidelberg (2011)
22. Yan, Z., Dijkman, R., Grefen, P.: Fast Business Process Similarity Search with Feature-Based Similarity Estimation. In: Meersman, R., Dillon, T.S., Herrero, P. (eds.) OTM 2010, Part I. LNCS, vol. 6426, pp. 60–77. Springer, Heidelberg (2010)
23. Jin, T., Wang, J., Wu, N., La Rosa, M., ter Hofstede, A.H.M.: Efficient and Accurate Retrieval of Business Process Models through Indexing. In: Meersman, R., Dillon, T.S., Herrero, P. (eds.) OTM 2010, Part I. LNCS, vol. 6426, pp. 402–409. Springer, Heidelberg (2010)
24. Dijkman, R.M., Vanderfeesten, I., Reijers, H.A.: The Road to a Business Process Architecture: An Overview of Approaches and their Use (2011)
25. Eid-Sabbagh, R.H.: Towards Automatic Generation of Process Architectures for Process Collections. In: 4th Central-European Workshop on Services and their Composition (ZEUS 2012) On-site Proceedings, pp. 88–95 (2012)
26. Rosa, M.L., Reijers, H.A., van der Aalst, W.M., Dijkman, R.M., Mendling, J., Dumas, M., Garcia-Banuelos, L.: Apromore: an advanced process model repository. Expert Systems with Applications 38(6), 7029–7040 (2011)
27. Hohpe, G., Woolf, B.: Enterprise Integration Patterns – Designing, Building, and Deploying Messaging Solutions. Addison-Wesley Longman Publishing Co., Inc., Boston (2003)
28. Witten, I.H., Frank, E., Hall, M.A.: Data Mining: Practical Machine Learning Tools and Techniques, 3rd edn. Morgan Kaufmann, Burlington (2011)

Aspect Oriented Business Process Modelling with Precedence

Amin Jalali[1], Petia Wohed[1], and Chun Ouyang[2,3]

[1] Department of Computer and Systems Sciences, Stockholm University, Sweden
{aj,petia}@dsv.su.se
[2] Science and Engineering Faculty, Queensland University of Technology, Australia
c.ouyang@qut.edu.au
[3] NICTA, Queensland Research Laboratory, Brisbane, Australia

Abstract. Complexity is a major concern which is aimed to be overcome by people through modelling. One way of reducing complexity is separation of concerns, e.g. separation of business process from applications. One sort of concerns are cross-cutting concerns i.e. concerns which are scattered and tangled through one or several models. In business process management, examples of such concerns are security and privacy policies. To deal with these cross-cutting concerns, the aspect orientated approach was introduced in the software development area and recently also in the business process management area. The work presented in this paper elaborates on aspect oriented process modelling. It extends earlier work by defining a mechanism for capturing multiple concerns and specifying a precedence order according to which they should be handled in a process. A formal syntax of the notation is presented precisely capturing the extended concepts and mechanisms. Finally, the relevance of the approach is demonstrated through a case study.

Keywords: Business Process Modelling, BPMN, Aspect Oriented, Separation of concerns.

1   Introduction

The interest to business process management has increased considerably during the last decade. BPMN is one of the most widely spread notation of business process modelling. Business processes are associated with a set of requirements some of which also reflect different concerns. Examples of concerns are security and logging. Concerns are typically *cross-cutting*, i.e. they are relevant for several business processes. For example, Figure 1 shows four typical concerns from the banking domain that spans across four processes. In addition, concerns can also be reflected in several places in one same process, i.e. they are *scattered* through a process.

Traditionally, as can be seen from Figure 1, the concerns are modelled as an integral part of the processes. This often leads to complex, inflexible and less reusable solutions. The complexity is increased as the number of tasks in a process grows to cover both business logic and cross-cutting concerns. The solution

J. Mendling and M. Weidlich (Eds.): BPMN 2012, LNBIP 125, pp. 23–37, 2012.

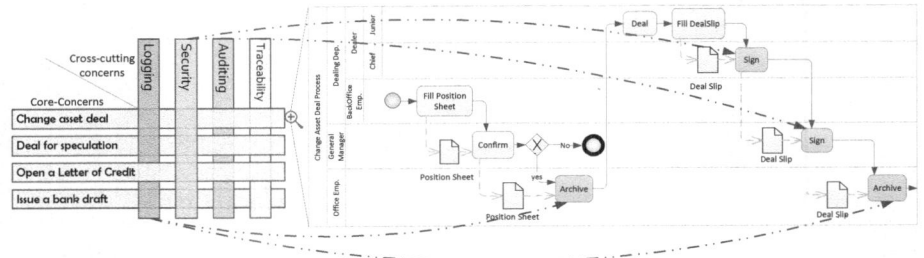

Fig. 1. Example of concerns in the context of a business process

is not flexible as changes in a concern have to be reflected in multiple places. Reusability is not supported due to the lack of placeholders for the concerns that can be refereed to when relevant.

To address these issues the aspect oriented principle has been proposed. In essence, this is a *separation of concerns*, advocating the separation of cross-cutting concerns from the core business process logic (which for short will be called core concerns). Within the programming paradigm, this principle is realised in Aspect Oriented Programming (AOP) (see for instance AspecJ [2]). In the business process management paradigm the aspect oriented principle has been introduced only recently. Charfi et al. [9] elaborate on how the separation of concerns can be handled in business process modelling by extending the Business Process Modelling Notation (BPMN) with notions for aspect oriented process modelling. They also extend BPEL [8] with features for aspect oriented Web service composition. Another existing effort is seen as the work by Cappelli et al. on proposing a different notation for aspect oriented business process modelling [1].

However, when applying the existing approaches, we recognised that not all the concerns could be separated from a business process model, due to the fact that none of these approaches can capture multiple concerns with sequential order of execution. In this paper we take the advances from [9, 8] and extend the approach presented in [9]. The contributions are three-fold. Firstly, we define a requirement which is necessary for capturing multiple concerns in a process with specific orders. We called it precedence requirement and extend the findings in [9] to fulfil this requirement. Secondly, we provide a rigorous formalisation of our approach to precisely capture the extended concepts and mechanism. Finally, we study and examine the relevance of the extended aspect oriented modelling mechanism using a case study.

The remainder of the paper is organized as follows. In Section 2 we present a conceptualisation of aspect oriented business process modelling. This includes a set of requirements for designing aspect oriented process modelling paradigm, the concepts for aspect orientation in business process modelling, and a formalisation of the correspondingly extended mechanism. Section 3 demonstrates the approach through a case study. Section 4 discusses the limitations of the current findings. Section 5 presents an overview of the related work in the area. Finally, Section 6 concludes the paper and presents directions for future work.

2   Approach

In order to provide support for aspect oriented business process modelling, some terminology need to be introduced. This terminology is influenced by the terminology in Aspect Oriented Programming. To exemplify it, we use the Business Process Modeling Notation (BPMN) [18]. We choose BPMN because: (i) it is a well known and widely spread out modelling notation and (ii) because our work initially targeted to extend the work by Charfi et al. [9] where BPMN was extended for the purposes of aspect oriented business process modelling. However, it should be noted that the conceptualization proposed here is general and could be adapted to extend other modelling notations such as UML Activity Diagrams, EPC, YAWL, etc. We start the presentation with a discussion of some basic requirements.

2.1   Requirements

When developing support for aspect oriented business process management there are some important requirements that need to be considered. These are compiled in [19] for the software engineering domain, but they are general and therefore applicable for the business process management domain as well. We summarize them in the list below and discuss their application in the business process management domain:

R1 It should be possible to identify and encapsulate concerns *simultaneously*. The concerns are *equal*, i.e. there is not a dominant concern that obstructs the extraction of other concerns. This means that *a notation supporting aspect oriented business process modelling shall allow for the presentation of multiple concerns relevant for a process*. In addition we identify the need for associating several concerns to one activity. This means that the notation *should be able to express the precedence order between multiple concerns* associated to an activity. I.e., it should be possible to specify both parallel and sequential order of execution of the concerns.

R2 It should be possible to identify and add concerns *incrementally* at any time during the development lifecycle. For business process management this means that *the addition of new concerns at a later stage of the development should be easy and without the need of invasive re-modelling*.

R3 Developers should not be required to know details of concerns that do not affect their particular activities. I.e. concerns should be "encapsulated" and business process analysts should be able to deal with one complexity at the time. In other words, *it should be possible to profile analysts*, i.e. business analysts of the core processes, business analysts of the security policies or archiving routines and etc.

R4 It must be possible to represent and manage *overlapping* and *interacting* concerns. For example, the Logging concern for a business entity may contain security elements. Therefore, in business process management *it should be possible to identify, model, execute and maintain processes which contain overlapping concerns*.

R5 "any separation of concerns mechanism must also include powerful integration mechanisms" [19]. In business process management context, it means that it is important to develop services (or software modules) that extend the behaviour of present workflow management systems in such a way that they can interpret and enact models that are produced with the aspect oriented principle.

While requirements R1-R4 are applicable in the design of an aspect oriented modelling notation, R5 is clearly related to the design of the underlying software. Hence for this paper, the first four requirements are of interest. In line with R5 we designed a service called the Aspect Service using Coloured Petri Nets (CPN) and present it in [13].

2.2   Concepts

We describe the concepts of aspect orientation with a fictitious `Transfer Money` process of a bank (see Figure 2). The process starts with a customer submitting a request of transferring money, i.e. `Fill form` activity. If the transfer is directed to an account owned by the customer, it is executed directly (i.e. activity `Transfer money`), if not the customer is asked to sign the transfer request (activity `Sign Transaction`), and then an automated `Detect fraud` activity is executed. After the money has been transferred, the transaction is archived (`Archive information` activity). If the transfer is made to an account with a different owner, the customer is also notified (`Notify Customer` activity), which is done before the archiving.

Looking closer into the `Transfer Money` process we can identify two concerns namely Security and Logging. The activities related to these concerns are coloured in two different ways to distinguish them from the core process. Figure 3 shows the same process modelled according to the aspect oriented principle. This implies that the Logging and the Security concerns are extracted from the core process and modelled as individual processes. We adopt the terminology introduced in aspect oriented programming and call the representation of concerns for *Aspects*. Although not shown in the example, an aspect can contain more than one process. These processes are called *Advices*. An advice contains a *PROCEED* activity, while a core process contains *Join Points*. The joint points show the possible places in a process where an aspect can be related to a process.

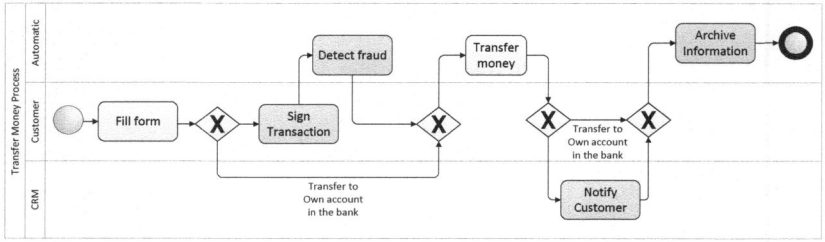

Fig. 2. Transfer Money Process - modelled in traditional way

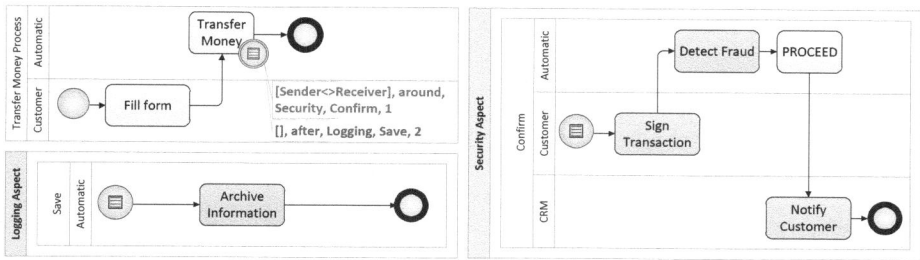

Fig. 3. Transfer Money Process - modelled according to the aspect oriented principle

For BPMN, these are all activities. When an activity is related to an advice, it is called *advised join point* activity. The `Transfer Money` activity in Figure 3 is an example of this. We propose the use of a *conditional event* on the border of an activity for indicating an advised join point.

In addition, each conditional event is annotated. The annotation shows the relevant advice for the joint point and the condition which needs to be fulfilled in order to trigger this advice. These conditions are called *pointcuts*. In the `Transfer Money` process, the `Confirm` advice is triggered only if a transfer request specifies a different account owner. Furthermore, the annotation shows when the advice process shall be executed in relation to the advised join point activity. The alternatives *before*, *after*, and *around* are possible. When the around alternative is specified, the corresponding advice also need to contain a *PROCEED* activity. The *PROCEED* activity is a placeholder which shows where in an advice process the corresponding advised join point activity should be executed. An advice process can have zero or one *PROCEED* activity: when zero the advice process is called *implicit*, otherwise *explicit*.

Encapsulation is achieved by modelling each concern as an advice process. Advices are grouped into Aspects based on their focus, e.g. Security and Logging. To support the execution of a process like the one in Figure 3, the functionality of a WfMS needs to be extended so that execution sequence specified in Figure 2 can be derived. The "integration" of an aspect to a process is called *weaving*. We also developed a service, called the Aspect Service, which specifies the semantics of the weaving [13].

It should be noted that the model in Figure 3 exposes the need for associating multiple advices to an activity. In the example, the advices shall be executed in a sequence, which means that it is important to be able to specify the execution order for them. This is done as part of the annotation. We call this order for the *precedence order*. The precedence requirement is recognized from the programming area, but have not been considered in previous work related to business process modelling. We recognised the relevance of this requirement through a case study.

2.3 Formalisation

We present a formalization of the syntax of BPMN covering the set of core elements depicted in Fig. 4. Based on that we then define the syntax of BPMN

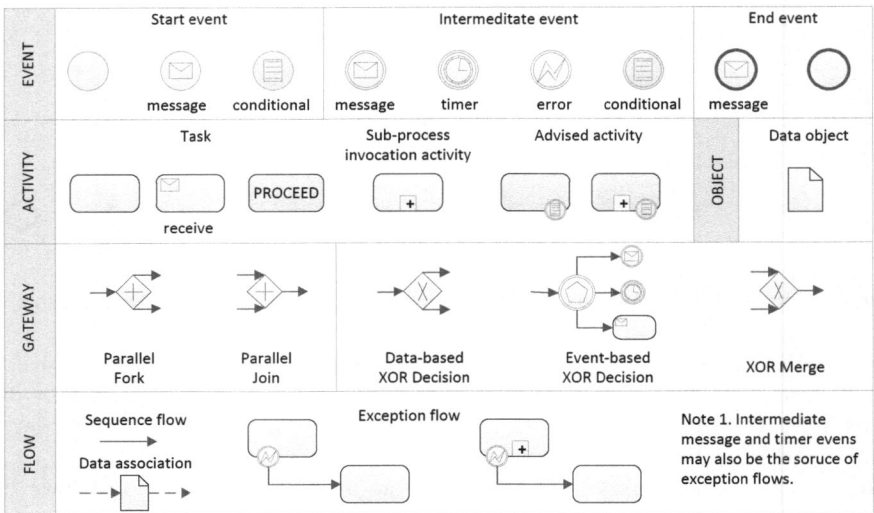

Fig. 4. A core subset of BPMN elements

extended with the *Aspects* considerations and refer to it as Aspect-Oriented Business Process Modeling Notation (AOBPMN).

The formalisation of the syntax of BPMN builds on the previous syntax definition (based on BPMN1.0) in [10] and extends it with the data and resource information and elaboration on events and exception constructs according to BPMN2.0 [18].

Definition 1 (Standalone BPMN Process). *A standalone BPMN process is a tuple* $\mathcal{M} = (\mathcal{O}, \mathcal{A}, \mathcal{A}^T, \mathcal{A}^S, \mathcal{E}, \mathcal{E}^S, \mathcal{E}^I, \mathcal{E}^E, \mathcal{G}, \mathcal{G}^A, \mathcal{G}^X, \mathcal{G}^E, \mathcal{G}^M, \mathcal{F}, \mathcal{D}, \mathcal{L}, \mathcal{R}, \mathcal{W}, EN,$ *Etype, Attch, Excp, Econd, Fcond, Dflw, Act, Belto, ADlab, Elab) where:*

- $\mathcal{O}$ *is a set of objects which can be partitioned into disjoint sets of activities* $\mathcal{A}$, *events* $\mathcal{E}$, *and gateways* $\mathcal{G}$,
- $\mathcal{A}$ *can be partitioned into disjoint sets of atomic activities (i.e. tasks)* $\mathcal{A}^T$ *and compound activities (i.e. subprocesses)* $\mathcal{A}^S$,
- $\mathcal{E}$ *can be partitioned into disjoint sets of start event* $\mathcal{E}^S$, *intermediate events* $\mathcal{E}^I$, *and end event* $\mathcal{E}^E$,
- $\mathcal{G}$ *can be partitioned into disjoint sets of parallel gateways* $\mathcal{G}^A$, *data-based exclusive decision gateways* $\mathcal{G}^X$, *event-based decision gateways* $\mathcal{G}^E$, *and exclusive merge gateways* $\mathcal{G}^M$,
- $\mathcal{F} \subseteq O \times O$ *is the control flow relation, i.e. a set of sequence flows connecting objects,*
- $\mathcal{D}$ *is a set of data objects associated with the process,*
- $\mathcal{L}$ *is a set of data object names,*
- $\mathcal{R}$ *is a set of roles designated to perform tasks or events within the process,*
- $\mathcal{W}$ *is a set of organisation groups involved in carrying out the process,*
- $EN = \{\textsf{message}, \textsf{timer}, \textsf{error}, \textsf{conditional}\}$ *is a set of basic event type names,*

- *Etype* : $\mathcal{E} \rightarrow \mathsf{EN}$ *is a function which assign to each event an event type,*
- *Attch* : $\mathcal{E}^I \nrightarrow \mathcal{A}$ *is a function*[1] *which attaches an intermediate event to an activity indicating that the event may occur during the activity execution,*
- *Excp* : $dom(Attch) \rightarrow \mathbb{B}$ *is a function*[2] *which specifies, for an intermediate event that is attached to an activity, whether or not its occurrence interrupts the (normal flow of) activity execution,*
- *Econd* : $\{e \in \mathcal{E} | Etype(e) = \mathsf{conditional}\} \rightarrow C$ *is a function*[3] *which assigns to each conditional event a condition specified as a boolean function,*
- *Fcond* : $\mathcal{F} \cap (\mathcal{G}^X \times O) \rightarrow C$ *is a function which maps sequence flows emanating from data-based exclusive decision gateways to conditions thus determining if the associated sequence flow is taken during the process execution,*
- *Dflw* : $\mathcal{D} \rightarrow (\mathcal{A} \cup \mathcal{E}^S \cup \mathcal{E}^I) \times (\mathcal{A} \cup \mathcal{E}^I \cup \mathcal{E}^E)$ *is a function which specifies the fact of each data object being transferred from one activity/event to another,*
- *Act* : $(\mathcal{A}^T \cup \mathcal{E}) \rightarrow 2^{\mathcal{R}}$ *is a function which designates one or multiple roles eligible to perform a task or event,*
- *Belto* : $\mathcal{D} \rightarrow \mathcal{W}$ *is a function which assigns a role to an organisation group.*
- *ADlab* : $\mathcal{A} \cup \mathcal{D} \rightarrow \mathcal{L}$ *is a function which labels each activity or data object,*
- *Elab* : $\mathcal{E} \nrightarrow \mathcal{L}$ *is a function which labels an event (without mandating that each event is labelled).*

For a standalone BPMN process $\mathcal{M}$, if ambiguity is possible, we use $\mathcal{M}$ as subscripts to each element defined in the tuple $\mathcal{M}$. For example, $\mathcal{A}^S_{\mathcal{M}}$ refers to the set of subprocess invocation activities in $\mathcal{M}$. Next, we define the syntax of a core BPMN process which supports a hierarchical structure comprising a set of standalone BPMN processes.

Definition 2 (Core BPMN Process). *A core BPMN process is a tuple* $\mathcal{P} = (\mathcal{Q}, \mathcal{M}^{top}, \mathcal{S}^{\diamond}, map, HR)$ *where:*

- $\mathcal{Q}$ *is a set of standalone BPMN processes,*
- $\mathcal{M}^{top} \in \mathcal{Q}$ *is the top level process,*
- $\mathcal{S}^{\diamond} = \bigcup_{\mathcal{M} \in \mathcal{Q}} \mathcal{A}^S_{\mathcal{M}}$ *is the set of all subprocess invocation activities in* $\mathcal{Q}$,
- $map : \mathcal{S}^{\diamond} \rightarrow \mathcal{Q} \backslash \{\mathcal{M}^{top}\}$ *is a function which maps each subprocess invocation activity to a standalone BPMN process, and*
- $HR = \{(\mathcal{M}, \mathcal{M}') \in \mathcal{Q} \times \mathcal{Q} \mid \exists_{s \in \mathcal{A}^S_{\mathcal{M}}} map(s) = \mathcal{M}'\}$ *is a connected graph,*

Next, an *advice* process is a BPMN process in which all the start events of the (top-level) process are conditional events and there may be one or more *PROCEED* activities, and an *aspect* process comprises a number of advice processes that belong to the same aspect.

Definition 3 (Advice Process). *An advice process* $\mathcal{P}^a = (\mathcal{Q}, \mathcal{M}^{top}, \mathcal{S}^{\diamond}, map, HR, \mathcal{A}^{Tp})$ *is a core BPMN process that satisfies the following conditions:*

---

[1] $\nrightarrow$ indicates a 'non total function', i.e. there are values in the domain that do not have a corresponding value in the range

[2] $\mathbb{B}$ is the boolean set {true, false}.

[3] C is the set of all possible conditions. A condition is a boolean function, operating over a set of propositional variables, which evaluates to true or false.

- $\forall e \in \mathcal{E}^S_{\mathcal{M}^{top}}$, $Etype(e) = $ *conditional, i.e. all start events in the top level process are conditional events, and*
- $\mathcal{A}^{Tp} = \bigcup_{\mathcal{M} \in \mathcal{Q}} \{a \in \mathcal{A}^T_\mathcal{M} | ADlab_\mathcal{M}(a) = PROCEED \wedge a \notin ran(Attch_\mathcal{M})\}$ *where PROCEED is a preserved label for PROCEED activities.*

Definition 4 (Aspect Process). *An aspect process is a tuple* $\mathcal{P}^A = (\{\mathcal{P}^a_1, \mathcal{P}^a_2, ..., \mathcal{P}^a_n\}, \mathcal{AN}, Advice)$ *where:*

- $\{\mathcal{P}^a_1, ..., \mathcal{P}^a_n\}$ *is a set of advice processes,*
- $\mathcal{AN}$ *is a set of advice names, and*
- *Advice* : $\{\mathcal{P}^a_1, ..., \mathcal{P}^a_n\} \to \mathcal{AN}$ *is a bijective function which assigns to each advice process a unique advice name.*

Finally, a AOBPMN process comprises a main BPMN process and a set of associated aspect processes. The interactions between the main process and the aspect processes, which are defined in the *pointcut* specifications, are carried out at the corresponding *advised join point activities*.

Definition 5 (Core AOBPMN Process). *A core AOBPMN process is a tuple* $\mathcal{AP} = (\mathcal{P}, \mathcal{E}_A, \mathcal{A}_{JP}, \{\mathcal{P}^A_1, \mathcal{P}^A_2, ..., \mathcal{P}^A_n\}, \mathcal{CN}, Aspect, Pointcut)$ *where*

- $\mathcal{P} = (\mathcal{Q}, \mathcal{M}^{top}, \mathcal{S}^\diamond, map, HR)$ *is a core BPMN process,*
- $\mathcal{E}_A = \bigcup_{\mathcal{M} \in \mathcal{Q}} \{e \in dom(Attch_\mathcal{M}) \mid Etype_\mathcal{M}(e) = $ *conditional* $\wedge Excp_\mathcal{M}(e)\}$ *is the set of intermediate conditional events attached to an activity in* $\mathcal{P}$,
- $\mathcal{A}_{JP} = \bigcup_{\mathcal{M} \in \mathcal{Q}} \{a \in \mathcal{A}_\mathcal{M} \mid \exists_{e \in \mathcal{E}_A} Attch_\mathcal{M}(e) = a\}$ *is a set of advised join point activities in* $\mathcal{P}$,
- $\{\mathcal{P}^A_1, ..., \mathcal{P}^A_n\}$ *is a set of aspect processes,*
- $\mathcal{CN}$ *is a set of aspect names,*
- *Aspect* : $\{\mathcal{P}^A_1, ..., \mathcal{P}^A_n\} \to \mathcal{CN}$ *is a bijective function which assigns to each aspect process a unique aspect name.*
- *Pointcut* : $\mathcal{E}_A \to 2^{Expr}$, *where* $Expr = \{\langle cond, pos, cn, an, order \rangle | cond \in C \wedge pos \in \{$ *before, after, around* $\} \wedge cn \in \mathcal{CN} \wedge an \in \mathcal{AN}_{Aspect^{-1}(cn)} \wedge order \in \mathbb{Z}^+\}$, *is a function which relates an event* $e \in \mathcal{E}_A$ *to a set of expressions, and each expression specifies for the corresponding advised join point* $a \in \mathcal{A}_{JP}$:
  * *the condition capturing the constrains for triggering an advice (cond),*
  * *when the advice should be triggered in relation to a (pos),*
  * *the aspect name (cn) and the advice name (an), and*
  * *the precedence order of triggering this advice among the multiple advices associated with a (order).*

3   Case Study

In this section, we apply the proposed approach on a real case study from the financial domain. The case study demonstrates how the approach can be applied for modularizing cross-cutting concerns in a banking process model. In particular, the case confirms the relevancy of the precedence requirement.

The banking case was selected due to previous knowledge in that domain. To choose appropriate processes, i.e. fairly simple yet representative processes with at least a couple of cross-cutting concerns, we conducted an interview with a domain expert from a bank. For confidentiality reason, the bank asked to be remained anonymous. Two processes were selected. Here, we present one of them namely the *Change asset deal* process[4]. Detailed information about the process was derived through a follow-up interview with the same domain expert.

Generally, the assets of the bank are in two forms, cash and non-cash. Cash assets are either in the form of the account balances of the bank or the marketable securities. The `Change asset deal` process (see Figure 5) handles deals for exchanging assets of the bank from one currency to another. The process starts with a *back office employee* filling in a position sheet (`Fill position sheet` activity). Then, the *general manager* either confirms or denies the deal. If the sheet is denied, the process ends. If the position sheet is approved, it is archived. Then, a *junior dealer* makes the deal and fills in a deal slip. Next, both a *chief dealer* and the *general manager* sign the deal slip, after which the deal slip is archived.

After the deal slip has been archived, two parallel sets of activities are performed. On the one hand, the dealt amount of money is sent to the external partner of the deal. For this, first an *employee of the Swift department* provides a swift draft for sending the money. Then, for security purposes, the *dealer*, *chief dealer* and *general manager* sign the swift draft. Finally, an *employee of the Swift department* sends out the swift. On the other hand, the dealt amount of money should be received. This part starts when an *employee of the Swift department* receives an NT300 swift message. The *employee* sends this message to the general manager. The *general manager* makes an order to the Back office department and to the dealer to control the swift message. These orders are issued separately. When each one of them has been controlled, the messages are archived (separately). When the deal is made, a *back office employee* registers a voucher in the accounting system. Finally, the deal is archived.

Figure 5 shows the models including both BPMN and AOBPMN versions of the change asset deal process. We distinguish the following results of applying the aspect oriented modularization approach:

- The aspect oriented solution *documents additional knowledge* of a business processes in the model. This knowledge specifies the relation between cross-cutting concerns and specific activities. For example, in Figure 5b, two security concerns are associated to the `Send Swift` activity; while, this knowledge is missed in the model in Figure 5a. I.e., it is not obvious to which of the two activities, `Provide Swift Draft` or `Send Swift`, the security concerns are related.
- The aspect oriented approach presented here enables the *separation of several concerns*. For instance in Figure 5b, two different aspects are associated to the `Fill DealSlip` activity. In this way, security policy makers could easily define and change their related policy without changing the main process or the archiving concerns.

---

[4] A detailed description and analysis of both process can be find in [12].

32   A. Jalali, P. Wohed, and C. Ouyang

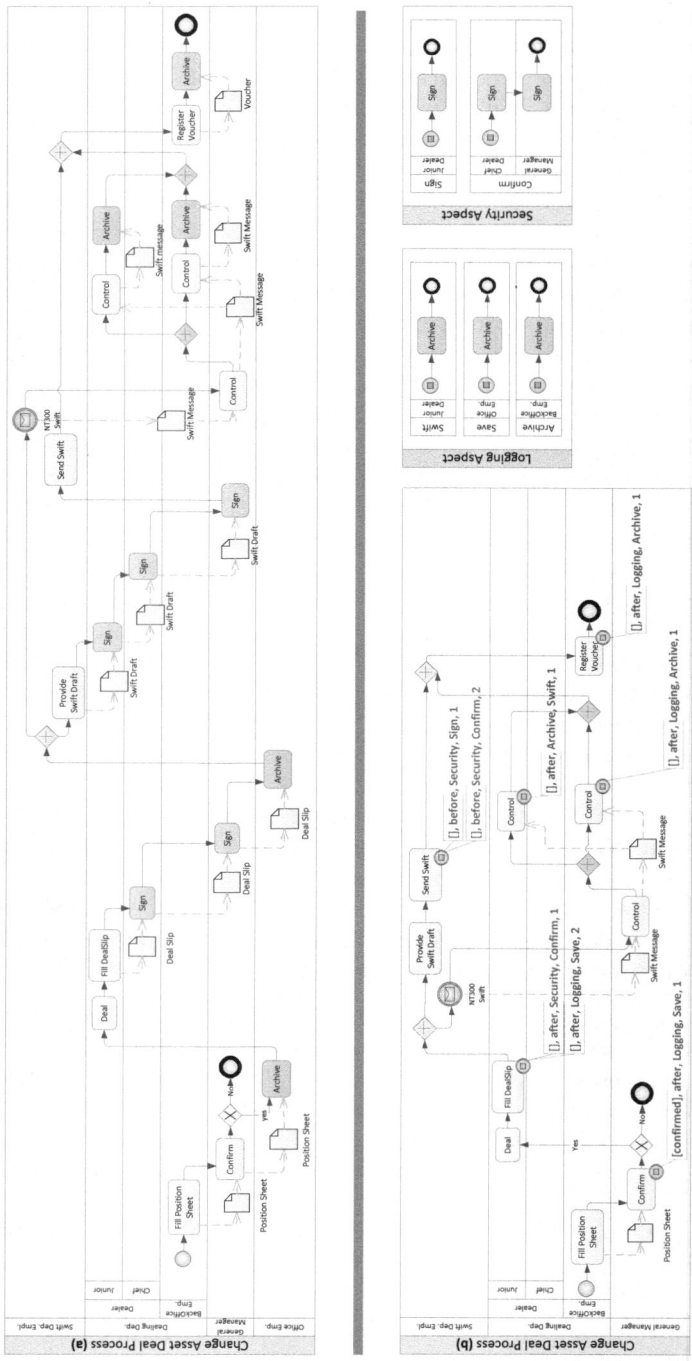

Fig. 5. The case study process: (a) traditional modelling; (b) AOBPMN modelling

– This approach enables separation of concerns which have different orders for consideration. For example, a dealslip should be first confirmed and then archived (see `Fill DealSlip` activity). Other approaches [9, 1] are not able to separate archive concern in this example, because they do not capture precedence requirement in the definition of advices. Therefore, they could not separate all concerns from a business process. In contract, our approach supports full degree of separation.

– The aspect oriented model in Figure 5b is *less complex*, in terms of number of activities, than the model in Figure 5a. While the model in Figure 5a contains 20 activities, the model in Figure 5b contains 10 activities in the main process and 6 in the advice processes. Hence, communicating the aspect oriented model to business users is expected to be easier [17].

– Aspect orientation *increases the reusability* since policies are defined once and can be used many times. See for instance the use of `Confirm` advice in Figure 5b, where it is associated both to `Fill DealSlip` and `Send Swift` activities in the core business process.

– It also *facilitates the maintenance* of the system. If a policy is changed, it should be applied in one model rather through all involved business processes. E.g., if the `Confirm` concern is changed, the updates are reflected in the corresponding advice in Figure 5b rather than on a number of places in a process and even in several processes (Figure 5a).

– Last but not least, aspect oriented modelling *enables agile development* of business processes, due to faster response to changes, better adaptability and flexibility [4, 3]. This enables incremental development of business processes i.e. the ability to add or change aspects also sometime after the development of the main process.

4 Limitations

During the work we encountered two types of limitations: limitations on the approach and limitations of the case study. We report on these here.

The *limitations of the case study* are implied by the characteristics of the two processes that the case study was run on, i.e. small size processes containing approximately 20 activities. Because, this was our first case study, we aimed at studying small real processes deeply. For this reason we did not look at too large processes. Instead we selected processes that we could learn quickly and that were suitable for presenting to a less domain knowledgeable audience. A side effect from this was that the advice processes we separated out were too small, i.e. several of them containing one activity only. While this raises the question whether it is meaningful to model and maintain processes with single activities, we believe that this phenomenon needs to be studied further. We would need to study how frequent this occurs and whether the benefits of separating the concerns outweigh the disadvantages of separating and maintaining small advice processes. Naturally, these will be questions for a follow-up case study where the applicability of the approach on bigger processes should be explored.

A *limitation of the approach* follows from the fact that we do not transfer information about resources among the core process and advices. This implies that we may have to define one advice multiple times if a resource who should perform an activity is different. For example, in the case study, three almost identical advices were defined to capture the Logging Aspect (see Figure 5b). If this limitation is addressed, the three processes in the Logging Aspect would be reduced to one with the corresponding resource configurations. On the other hand, this resource limitation may not be a trivial thing when several resources are included in the execution of an advice.

Finally, while the formalisation of the syntax for aspect oriented business process modelling is presented in this paper, the semantics is formalised through Coloured Petri Nets and presented in [13]. As the work was carried out iteratively, the precedence requirement was confirmed through the case study and captured in the formal syntax of the approach. However, it also needs to be reflected in the formal semantics in future. The limitations outlined here present some lines for future work.

5   Related Work

The work on aspect oriented business process management is inspired from two fields, namely *software development* and *requirements engineering* that have been carried out primarily by the research groups of Charfi and Capelli, correspondingly. Within the software development, a lot of work has been done in the area of Aspect Oriented Programming(AOP) e.g., [15, 2, 16, 14, 11]. The ideas from AOP were initially transferred to the web services composition domain. Recently, they were also utilized in BPM area.

Charfi, et al. extended BPEL to support Aspect-Oriented Web Service Composition [6–8], and called the extension AO4BPEL. Although this work does not address people involvement in processes, it opened up further investigation of aspect orientation for process composition. For example, an extension of BPMN was proposed to support aspect oriented business process modelling, which is referred to as AO4BPMN [9]. Based on the terminology from AOP, AO4BPMN defines the notions for aspect, advice, pointcut, join point and proceed. AO4BPMN enables the modelling of aspects and advices through decomposition (i.e. in separate pools and swimlanes) and uses annotations on the join point activities, for relating the advices to the main process. We founded our work on the approach by Charfi et al. [9] and provide the following contributions. First, we fine-tune the application of aspect orientation terminology on BPMN. E.g. annotation (which were used to relate the advices with the main process) are replaced with intermediate conditional event, as these actually affect the flow of a Process [18]. By using the notion of intermediate conditional event, we ensure compliance with the BPMN specification [18]. Second, we provide a formalization of the syntax. Third, from the case study we identified the need for specifying Precedence between advices associated to the same joint point activity [12] and included mechanisms for expressing these in the formalization.

The other approach, by Cappelli, et al., introduces a notion for aspect oriented BPM [5, 1, 20, 21]. The work was initiated with a conceptualization of the terms used for aspect oriented process modelling language. It is based on BPMN and implemented in the CrossOryx editor. The approach [1] also outlines criteria for identifying cross-cutting concerns. This work was demonstrated with the application of aspect oriented business process modelling in a case. It also develops a conceptual model for Aspect Oriented Process Modelling Language (AOPML) [5]. It also describes criteria for identifying cross-cutting concerns. In addition, an extension of BPMN with aspect oriented notions and a representation language for pointcut specification [1] were defined and implemented. Later on the approach was adapted for relating aspect oriented business process models with goal [21] and service identification [20]. From this body of work, [1] is the most relevant to our work. In contrast to [1] we (i) do not add new notation to the BPMN standard, but use the existing elements for capturing aspect oriented models; (ii) we deal with precedence; and (iii) we remove ambiguity by providing formalization of the syntax. E.g., in [1] the application of "around" (called "during") is not clarified. In addition to the work by Charfi, et al. [9] and Cappelli, et al. [1], we also provide a formal semantics to our approach, which is presented in [13].

6 Conclusions

In this work, we elaborated on aspect oriented business process modelling and demonstrated its application for reducing the complexity of process models. We also outlined the need of precedence requirement, i.e. the need for specifying the execution order of multiple advices associated to the same activity. As an outcome, we extended the AO4BPMN approach [9] to support the definition of precedence. We also modified the approach to comply with the BPMN specification and presented a formal syntax. We validated the approach using a real banking case study, which illustrated how the approach separates concerns, prioritises the handling of concerns, reduces model complexity, increases reusability, documents additional domain knowledge, enables agile process development, increases flexibility and facilitates the maintenance of process models. In a sequel paper [13] we specify the formal semantics of the approach.

During our work, we reflected on a number of requirements presented for the software development domain. Our approach fulfils these requirements. It allows for the *modelling of multiple advices* associated to a process as well as multiple advices associated to a single activity. Aspects can be overlapping, i.e., *advices can be related to each other*. For example, an advice can contain activities associated to another advice. Finally, due to encapsulation, the definition of advices can be done *incrementally* and carried out by *different stakeholders*.

Some directions for future work include extending the semantics of the weaving mechanism to deal with precedence. Moreover, the approach shall be extended with a mechanism for passing resource information to advices. Finally, it is also desirable to carry out a more extensive case study to explore the benefits of aspect oriented business process modelling of larger processes and in other domains such as health care and the public sector.

Acknowledgements. This work is partially supported by National ICT Australia (NICTA). NICTA is funded by the Australian Government as represented by the Department of Broadband, Communications and the Digital Economy and the Australian Research Council through the ICT Centre of Excellence program.

References

1. Cappelli, C., et al.: Reflections on the modularity of business process models: The case for introducing the aspect-oriented paradigm. Business Process Management Journal 16, 662–687 (2010)
2. Belblidia, N., Debbabi, M.: Formalizing AspectJ Weaving for Static Pointcuts. In: SEFM, pp. 50–59. IEEE Computer Society (2006)
3. Booth, R.: Agile manufacturing [management]. Engineering Management Journal 6(2), 105–112 (1996)
4. Burgess, T.F.: Making the Leap to Agility: Defining and Achieving Agile Manufacturing through Business Process Redesign and Business Network Redesign. International Journal of Operations and Production Management 14(11), 23–34 (1994)
5. Cappelli, C., Leite, J.C.S.P., Batista, T., Silva, L.: An aspect-oriented approach to business process modeling. In: Proceedings of the 15th Workshop on Early Aspects, EA 2009, pp. 7–12. ACM, New York (2009)
6. Charfi, A., Mezini, M.: Aspect-Oriented Web Service Composition with AO4BPEL. In: Zhang, L.-J., Jeckle, M. (eds.) ECOWS 2004. LNCS, vol. 3250, pp. 168–182. Springer, Heidelberg (2004)
7. Charfi, A., Mezini, M.: Hybrid web service composition: business processes meet business rules. In: Aiello, M., Aoyama, M., Curbera, F., Papazoglou, M.P. (eds.) ICSOC, pp. 30–38. ACM (2004)
8. Charfi, A., Mezini, M.: AO4BPEL: An Aspect-oriented Extension to BPEL. In: World Wide Web, pp. 309–344 (2007)
9. Charfi, A., Müller, H., Mezini, M.: Aspect-Oriented Business Process Modeling with AO4BPMN. In: Kühne, T., Selic, B., Gervais, M.-P., Terrier, F. (eds.) ECMFA 2010. LNCS, vol. 6138, pp. 48–61. Springer, Heidelberg (2010)
10. Dijkman, R.M., Dumas, M., Ouyang, C.: Semantics and analysis of business process models in BPMN. Information & Software Technology 50(12), 1281–1294 (2008)
11. Ho, W.-M., Jézéquel, J.-M., Pennaneac'h, F., Plouzeau, N.: A Toolkit for Weaving Aspect Oriented UML Designs. In: AOSD, pp. 99–105 (2002)
12. Jalali, A.: Foundation of Aspect Oriented Business Process Management. Master's thesis, Stockholm University (2011)
13. Jalali, A., Wohed, P., Ouyang, C.: Dynamic Weaving of Aspects for Business Process Management Systems. Technical report, Dept. of Computer and Systems Sciences, Stockholm University (March 2012)
14. Jézéquel, J.-M.: Model Driven Design and Aspect Weaving. Software and System Modeling 7(2), 209–218 (2008)

15. Kiczales, G., Lamping, J., Mendhekar, A., Maeda, C., Lopes, C., Loingtier, J.-M., Irwin, J.: Aspect-oriented Programming. In: Aksit, M., Matsuoka, S. (eds.) ECOOP 1997. LNCS, vol. 1241, pp. 220–242. Springer, Heidelberg (1997)
16. Klein, J., Fleurey, F., Jézéquel, J.-M.: Weaving Multiple Aspects in Sequence Diagrams. Transactions on Aspect-Oriented Software Development 3, 167–199 (2007)
17. Mendling, J., Reijers, H.A., Cardoso, J.: What Makes Process Models Understandable? In: Alonso, G., Dadam, P., Rosemann, M. (eds.) BPM 2007. LNCS, vol. 4714, pp. 48–63. Springer, Heidelberg (2007)
18. OMG. Business Process Model and Notation (BPMN), Version 2.0 (2011), http://www.omg.org/spec/BPMN/2.0/PDF/ (accessed March 2012)
19. Ossher, H., Tarr, P.: Multi-Dimensional Separation of Concerns and the Hyperspace Approach. In: Aksit, M. (ed.) Software Architectures and Component Technology, vol. 648, pp. 293–323. Springer, US (2002)
20. Perin-Souza, A., Cappelli, C., Santoro, F.M., Azevedo, L.G., do Prado Leite, J.C.S., Batista, T.V.: Service identification in aspect-oriented business process models. In: Gao, J.Z., Lu, X., Younas, M., Zhu, H. (eds.) SOSE, pp. 164–174. IEEE (2011)
21. Santos, N., Jack, F., do Prado Leite, S., Cesar, J., Cappelli, C., Batista, T.V., Santoro, F.M.: Using Goals to Identify Aspects in Business Process Models. In: Proc. of the 2011 Int. Workshop on Early Aspects, EA 2011, pp. 19–23. ACM, New York (2011)

BPMN4TOSCA: A Domain-Specific Language to Model Management Plans for Composite Applications

Oliver Kopp, Tobias Binz, Uwe Breitenbücher, and Frank Leymann

Institute of Architecture of Application Systems,
University of Stuttgart,
Universitätsstraße 38, 70569 Stuttgart, Germany
{kopp,binz,breitenbuecher,leymann}@iaas.uni-stuttgart.de

Abstract. TOSCA is an upcoming standard to capture cloud application topologies and their management in a portable way. Management aspects include provisioning, operation and deprovisioning of an application. Management plans capture these aspects in workflows. BPMN 2.0 as general-purpose language can be used to model these workflows. There is, however, no tailored support for management plans in BPMN. This paper analyzes TOSCA with the focus on requirements on workflow modeling languages to come up with a strong link to the application topology with the goal to improve modeling support. To simplify the modeling of management plans, we introduce BPMN4TOSCA, which extends BPMN with four TOSCA-specific elements: TOSCA Topology Management Task, TOSCA Node Management Task, TOSCA Script Task, and TOSCA Data Object. Portability is ensured by a transformation of BPMN4TOSCA to plain BPMN. A prototypical modeling tool supports the strong link between the management plan and the TOSCA topology.

Keywords: Cloud Computing, Service Management, Management Plans, BPMN Extension.

1 Introduction

To decrease cost and prevent vendor lock-in, portability of applications is—especially in the area of cloud computing—very important. To face this challenge, the *OASIS Topology and Orchestration Specification for Cloud Applications* (TOSCA) [1] is a way to describe the structure of portable services in a topology and their management as workflows, so called plans. A topology consists of node templates which offer management operations to create new instances or deploy software artifacts, for instance. Currently, the BPMN management plans directly point to the service interfaces and are not linked to the topology anymore. Therefore, we propose BPMN4TOSCA, a domain-specific BPMN [2] extension, which enables convenient integration and direct access to the TOSCA topology and provided management operations.

J. Mendling and M. Weidlich (Eds.): BPMN 2012, LNBIP 125, pp. 38–52, 2012.

Our contribution is fourfold: (i) Analyzing the requirements for modeling TOSCA management plans using BPMN, (ii) the BPMN extension BPMN4TOSCA allowing tight integration of topology data and management operations into plans, (iii) a transformation of BPMN4TOSCA into standard-compliant BPMN, and (iv) a prototypically implemented BPMN4TOSCA support in a TOSCA modeling tool.

The paper starts with a general introduction to the concepts behind TOSCA: the topology templates and the management plans (Sect. 2). Section 3 presents a concrete TOSCA use case, where the concepts of TOSCA are detailed. Based on this use case, general requirements on the plan modeling language are derived in Sect. 4. Based on the requirements, Sect. 5 presents BPMN4TOSCA, a domain-specific variant of BPMN supporting TOSCA management plan modeling. As typical workflow engines are not capable of executing extended BPMN, we present in Sect. 6 how to transform BPMN4TOSCA to plain BPMN 2.0 to enable execution on standard workflow engines. Section 7 presents a prototype supporting modeling TOSCA documents including management plans expressed in BPMN4TOSCA. Subsequently, Sect. 8 surveys on related work including the field of modeling composite applications and service management. Finally, Sect. 9 concludes and presents an outlook on future work.

2   Fundamentals

The *Topology and Orchestration Specification for Cloud Applications*, TOSCA for short, is an exchange format to describe the components of composite applications, their relations, as well as how to manage them. TOSCA is currently standardized in an OASIS Technical Committee[1]. Its main goal is enabling portability of composite applications between different cloud management environments to prevent vendor lock-in and increase automation in service management. To facilitate this, a *service template* is described in TOSCA, as denoted in Fig. 1. It consists of two major parts: the service's topology and management plans. The *topology* captures the structure of the composite application as a graph of node templates which are semantically connected by relationship templates. Each template is of a certain type. The type defines its properties, lifecycle states, policies, related artifacts, and management operations. Types in TOSCA are extensible, i. e., they can be defined as part of the service template and are not a predefined closed set. *Deployment artifacts* attached to a node define how this node is implemented. For instance, a virtual machine image may be a deployment artifact for the Linux node type or a Java Web archive for the Web application node type. The management operations supported by a node, for example, start, backup, or upgrade a node, are defined as WSDL Web service, REST service, script, or a combination thereof. If a management operation is not provided by the deployment artifact itself, e. g., an application server offering an JMX management service, or an external service, i. e., Amazon EC2 to start up virtual machines in their cloud, it can be included inside the service

---

[1] http://www.oasis-open.org/committees/tosca

template as so called *implementation artifact*. This enables service creators to ship management and administration services as part of their service. All in all, a service template consists of node templates and relationship templates. Each node template has a node type and a relationship template has a relationship type. A service template is instantiated to a service instance, where the node templates become nodes and the relationship templates become relationships.

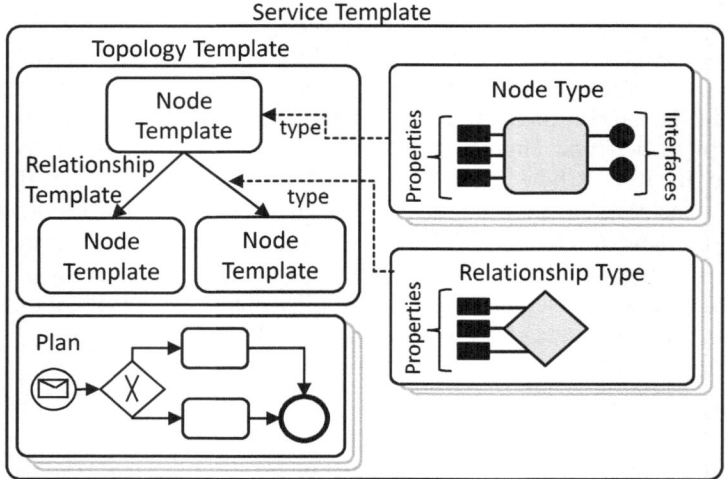

Fig. 1. Overview of TOSCA Building Blocks (adapted from [1])

TOSCA enables service creators to model the management aspects of services into plans. Plans express higher-level management tasks, which are, for example, how to setup the service, how to scale it up, back it up, or upgrade all operating systems. Having the management explicitly in the service template makes the management knowledge portable, reusable, and enables automation. Plans are modeled by the developer of the application or experienced operators ensuring widespread usage of their accumulated best practice knowledge and relieves enterprise IT from some of the management burden. Plans orchestrate the different management operations offered by the nodes to fulfill their task. The TOSCA specification defines three types of management plans: Build, modification, and termination plans. Technically, plans are workflows written, for example, in BPMN [2] or BPEL 2.0 [3]. Binz et al. [4] discuss the advantages of using workflow technology for plans. The key benefits are fault handling, compensation, auditing, parallelism, and integration of humans.

TOSCA requires a compliant management environment to run the service templates. We call such an environment "TOSCA container". After importing a new service template, the TOSCA container ensures, for example, that the implementation artifacts implementing the management operations are available before the service is instantiated for the first time. Additionally, the container's responsibilities are to manage service templates and their instances, offer access to the topology model and instance data, and handle the deployment artifacts accompanying the service template. Before deploying the plans on a plan engine, the TOSCA container binds the plans to the respective endpoints. This is necessary as it is not known where the management operations have been deployed and in which management environment the service template will be executed. This binding is key to enable portable service management between different TOSCA containers. In this paper, we focus on the mechanisms required to use management operations and TOSCA container services in management plans.

TOSCA recommends BPMN 2.0 as workflow language to model management plans. Other workflow languages—such as BPEL 2.0—may also be used. In contrast to BPEL 2.0, BPMN 2.0 is currently the preferred choice as it offers a standardized graphical rendering and does not force the workflow graph to be acyclic [5]. Starting in version 2.0, BPMN has also a well-defined execution semantics. Tasks and the control flow between them are the central elements of BPMN determining what the workflow does and in which order. BPMN defines tasks to call services (*service task*), to execute scripts (*script task*), to trigger human actions (*human task*), and others which are not important in our context. In contrast to BPEL, data flow in BPMN is explicitly modeled by using data objects. Tasks and events read from and write to data objects by using data associations which may contain data transformation rules.

3   Use Case

In this section we describe a TOSCA use case used in the following section to derive the requirements towards BPMN4TOSCA. The use case describes an online bookstore whose architecture is presented in Fig. 2. Each component is rendered as a box representing a TOSCA node template. The dashed arrows denote relationship templates of type "hosted-on". The application uses Java Servlet and Java Server Pages technology and is packaged into a single WAR (Java Web archive) file. To run the application, the WAR file has to be deployed on a servlet container, *Apache Tomcat* in our case. The servlet container is hosted on an operating system, *Ubuntu 12.04 LTS* in the use case, which is hosted on a Amazon EC2[2] virtual server.

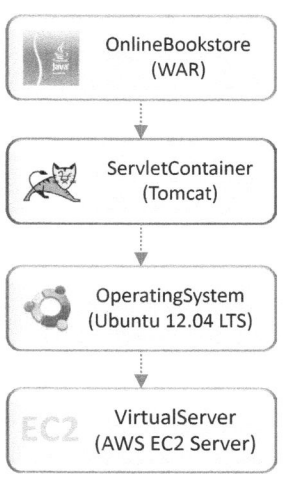

Fig. 2. Online Bookstore

In the following, we describe how to use the concepts of TOSCA to deploy and manage the online bookstore application without BPMN4TOSCA to show the limitations and inconveniences. To deploy the online bookstore, its components virtual server,

---

[2] http://aws.amazon.com/ec2/

operating system, servlet container, and the online bookstore application itself have to be deployed in the right order, typically from bottom up. The build plan orchestrates the management operations, scripts, and operations offered by the TOSCA container. The first step of the build plan in our use case is to create and start the virtual server using the operating system image defined in the topology, which establishes the hosted-on relationship defined in Fig. 2. Afterwards, the build plan invokes a management operation of the operating system to copy a bash script[3] onto the operating system. By invoking another management operation the script is invoked and installs the Tomcat servlet container on the operating system. This script uses Ubuntu's package management system to install Tomcat. After Tomcat is installed on the operating system, the servlet container is started by calling the operation `startService` offered by the operating system implementation artifact. The last step of building the application is deploying the online bookstore application on Tomcat. Therefore, the build plan invokes the `deployWar` operation implemented by the implementation artifact of the Tomcat node type and passes a reference to the WAR file. The operation deploys this WAR file—which also may be stored online—into Tomcat. All in all, the whole application is now deployed, running, and can be used.

4   Requirements

Based on the use case, we identified three requirements towards a solution which facilitates the tight integration of the management plans with the managed application topology.

TOSCA management plans typically process and manipulate properties of nodes and relationships, for example, the IP address of the virtual server node in our use case. In order to access and modify these instance properties, BPMN service tasks are used to call the respective TOSCA container APIs. Due to the fact that properties play a central role in management plans and, therefore, are heavily accessed by different tasks, the management plans get polluted with BPMN tasks. Thus, the management plans become complex and hard to understand. Business process research has shown that the maintainability and understandability of business processes decrease rapidly with the number of tasks. For example, Cardoso [6] proposes a measure for control-flow complexity increasing with the number of tasks and Reijers et al. [7] discuss the business process complexity in the context of modularization which is used to reduce the number of tasks. Therefore, requirement R1 for BPMN4TOSCA is reducing this complexity by providing ways to access and modify properties of nodes and relationships without modeling overhead in terms of business process elements.

In addition to R1, management plans must be able to access the TOSCA service topology model which is provided by the TOSCA container. TOSCA model access is required for dynamic plans, for example, to retrieve all nodes of a certain type. Therefore, requirement R2 is to enable the management plans to access the TOSCA topology model.

---

[3] http://www.gnu.org/software/bash/manual/bashref.html

To deploy, instantiate, and manage topologies, plans typically invoke management operations offered by nodes. The number of available management operations offered by nodes may become unmanageable for the modeler when there are many different nodes in the topology. Furthermore, as many operations will have similar names, e.g., `deploy` is a common operation name in this domain, and as operations may be spread across multiple TOSCA files, it becomes a complex task for the modeler to select the right management operation of the correct node. Thus, requirement R3 for BPMN4TOSCA is to ease the selection of management operations and to provide support for strong integration of management plans and management operations offered by nodes.

Scripts play an important role in the management of composite applications, especially during their deployment. They are widely used to perform installation and configuration tasks in systems management. Typically, these scripts are copied to the respective nodes and executed locally. TOSCA supports this concept by providing script operations which are attached to nodes. Hence, requirement R4 is: BPMN4TOSCA must support an easy and comfortable way to execute scripts on nodes.

5   BPMN4TOSCA: Enabling TOSCA Plan Modeling

In this section we introduce the BPMN language extension BPMN4TOSCA. To meet the requirements stated in the previous section, the design of BPMN4TOSCA consists conceptually of two parts: The first part provides a BPMN language extension (this section) and corresponding processing model, which defines the semantics of the extension (Sect. 6). The second part defines additional functionalities, which have to be provided by the modeling tool in addition to the language extension to provide the functionality (Sect. 7). Although the second part makes BPMN4TOSCA to more than an extension of BPMN, we nevertheless call the BPMN language extension BPMN4TOSCA, too.

The language extension consists of four new BPMN4TOSCA-elements, each accompanied with a graphical representation: TOSCA Topology Management Task (Sect. 5.1), TOSCA Node Management Task (Sect. 5.2), TOSCA Script Task (Sect. 5.3), and TOSCA Data Object (Sect. 5.4).

5.1   TOSCA Topology Management Task

The TOSCA Topology Management Task extends the BPMN service task in a way that standardized topology management operations offered by the container are predefined and can be directly used. An example operation is `getServiceTemplate` to get the TOSCA service template the plan works on. The selected operation is put in the attribute `operationRef` [2, p. 159]. This addresses requirement R2 of Sect. 4: By using a Topology Management Task, the operation can be directly chosen.

5.2   TOSCA Node Management Task

The TOSCA Node Management Tasks simplifies selecting and invoking management operations of nodes. It extends the BPMN Service Task. The node template id to work on is stored in the existing attribute `implementationRef` [2, p. 106]. The id itself is contained in the namespace of the service template. The selected operation is put in the attribute `operationRef`. This fulfills requirement R3: By using a Node Management Task, the user directly selects the node template to work on and the operation to call.

5.3   TOSCA Script Task

The TOSCA Script Task meets requirement R4 of Sect. 4 by providing the opportunity of referencing scripts and corresponding nodes on which they shall be performed. The TOSCA Script Task offers two possibilities to define scripts which should be performed on nodes: First, scripts can be defined inline the task itself, i. e., the script is part of the task description. Second, the TOSCA Script Task can reference scripts defined in TOSCA files as they are identified by unique ids. In addition to that, a TOSCA Script Task specifies the node on which the intended script has to be performed by using a unique id referencing to the corresponding node template defined in the TOSCA file.

The scripts must be able to be copied automatically to the nodes and executed on them. This is managed by the TOSCA container. For that, the container needs special operations provided by the nodes to enable this kind of generic script handling: Each of the target nodes has to provide an implementation artifact implementing an interface defining a set of pre-defined management operations prescribed by BPMN4TOSCA. The implementation artifact does the transformation from the generic container operation to the specifics of the respective scripting language, e. g., transforming parameter data types and setting environment variables.

The interface defines three main operations: `deployScript`, `runScript`, and `undeployScript`. `DeployScript` gets the actual script passed as parameter and returns a unique id which identifies the deployed script. `RunScript` gets this id and the input parameters for the script passed as parameter and returns the result of the script execution. `UndeployScript` also gets the id of the deployed script and undeploys it from the respective node.

The TOSCA Script Task inherits from BPMN's script task. In case the script is part of the task, the script task semantics and its attributes `scriptFormat` and `script` is re-used. In case the task references a script stored in the service template, these two attributes are not used. Instead three attributes are added: `scriptReference`, which references a script defined (or referenced) in the TOSCA file and `targetNodeTemplateId`, which references the node template on which the script has to be executed, `targetNodeInstanceId`, defining the concrete instance of the node template if multiple instances are allowed.

5.4 TOSCA Data Object

 The language extension meets requirement R1 of Sect. 4 by introducing a *TOSCA Data Object (TDO)*, which automatically provides access to runtime property information of nodes and relationships. TOSCA Data Objects provide information without the need to explicitly model BPMN service tasks requesting the respective information from the TOSCA container and sending modifications to the container. The data handling between TOSCA Data Object and TOSCA container is done automatically "behind the scenes" and invisible to the plan modeler. As soon as a TOSCA Data Object is defined in a plan, the referenced information is accessible and can be modified. The issue of dealing with multiple different plans using TOSCA Data Objects representing the same nodes or relationship properties concurrently lead to well-known transactional problems such as lost update or dirty reads breaking ACID properties [8]. This concurrent access to any information objects defined in the topology through different plans is a general problem, thus our approach does not need to deal with this issue as we assume that the TOSCA container is responsible for avoiding the concurrent execution of plans accessing the same information objects concurrently, i.e., the TOSCA container is responsible for plan scheduling. Plan scheduling denotes the order in which the plans are executed and not a refinement of the plans itself.

TOSCA Data Objects extends the BPMN data object by adding two TOSCA-related attributes to identify the nodes resp. relationships the TOSCA Data Object is reffering to: (i) A reference to the corresponding node template or relationship template whose properties should be reflected by the TOSCA Data Object, named `nodeTemplateId` and `relationshipTemplateId`, respectively. (ii) The optional attributes `nodeInstanceId` and `relationshipInstanceId` identify the concrete node instance or relationship instance if there are multiple instances, as defined by the min and max instances attributes in TOSCA. In case the data object is a collection, only the template id attribute is allowed. Iterating over each referenced node (or relationship) instance is enabled by the `inputDataItem` property of a looping BPMN task [2, p. 192].

6 Processing BPMN4TOSCA

The BPMN4TOSCA extension leads to a non-standards-compliant BPMN and, therefore, needs special treatment. The presented extensions are run-time extensions: They introduce new functionalities which are not natively supported by the executing environment. Nevertheless, there are two options to enable the extensions during runtime: (A1) extend the modeling tool *and* the workflow engine to support the new functionality and (A2) transform the functionalities into standards-compliant executable elements before deployment [9]. When choosing option A1, the workflow engine is extended to support the new functionality. The drawback is that the extension is supported by standards-compliant engines only if they also support the extension. To avoid requiring an implementation of the extension at workflow engines, one may provide a transformation of an extended process model to a standard process model (option A2). This generated model typically depends on external services to offer the functionalities.

In the case of BPMN4TOSCA, the transformation to plain BPMN is possible. We opted for a transformation instead of extending the workflow engine for the following reasons:

- Extending the execution environment for new capabilities renders TOSCA non-portable, because only standards-compliant BPMN is portable across different execution environments.
- Providing a modeling tool supporting the BPMN4TOSCA approach enables transformation of BPMN4TOSCA to plain BPMN by the export functionality of the modeling tool. Thus, BPMN4TOSCA is only visible in the tool and transparent to the execution environment as the semantics and functionalities remain only implicitly in the exported files while keeping all benefits of the approach for the modeler.

The following subsections show how the four BPMN4TOSCA tasks are transformed into standards-compliant BPMN:

6.1   TOSCA Topology Management Tasks

The TOSCA Topology Management Task references operations provided by the TOSCA container. A new implementation reference is added. It points to the concrete port type, where the WSDL service of the TOSCA container is offered. This indirection is necessary as the management operation interface is not standardized in TOSCA.

6.2   Transformation of TOSCA Node Management Task

The TOSCA Node Management Task references a node template in a service topology and one operation. This information is replaced by a reference to the concrete WSDL port type and WSDL operation of the referenced operation.

6.3   Transformation of TOSCA Script Tasks

TOSCA Script Tasks need to be transformed into BPMN service tasks as conceptually shown in Sect. 5.3. A Script Task references the script and the node on which it has to be performed. During plan transformation, the TOSCA Script Task is replaced by three BPMN service tasks which are executed in sequence: deploy script, run script, undeploy script. The TOSCA container binds these three tasks to the services offered by the implementation artifact of the corresponding node. These services implement the interface introduced in Sect. 5.3.

To deploy the script on the target node, the first task invokes the `deployScript` operation offered by the implementation artifact's service and passes the entire script and the type of the script, e. g., Ant script, to the service. The type of the script is required for selecting the corresponding script handlers, which define how each script type is processed. This is completely transparent to the plan. The required logic is implemented either in the implementation artifact or the node itself. For some script types, there possibly is a need for some kinds of agents installed directly on the node to interact with the implementation artifact, others are able to implement the whole script handling in the implementation artifact and the node remains totally unaware of executing scripts. This is script type specific implementation design and therefore out of scope: BPMN4TOSCA defines only the interfaces, not how to deal with scripts. Thus, for each script type, there is a specific implementation, i. e., a script handler, for dealing with this type. Referenced scripts have to be resolved by the TOSCA container before passing the script to the service. After successful deployment, the `deployScript` operation returns a unique identifier used to identify the deployed script for further steps, i. e., execution and undeployment of the script.

The second BPMN service task invokes the `runScript` operation and passes the script identifier and corresponding input parameters which are contained in the input data object associated with the TOSCA Script Task to the service. The implementation artifact uses the identifier to find the deployed script handler, passes the input parameters, and executes it. After successful execution, the operation returns the output parameter which is in turn returned from the script via its script handler and writes the values to the associated output data object. Passing data from data objects to data input and from data output is done using the BPMN way using `dataInputAssociation` and `dataOutputAssociation` [2, p. 224]. Defining these associations is out of scope and has to be done by the modeler.

The third BPMN service task invokes the `undeployScript` operation and passes the script identifier. The operation undeploys the executed script.

6.4   Transformation of TOSCA Data Objects

The transfer of property information data from the TOSCA container to plans happens transparently to the modeler. TOSCA Data Objects are converted to BPMN data objects. For reading TOSCA data, additional BPMN service tasks

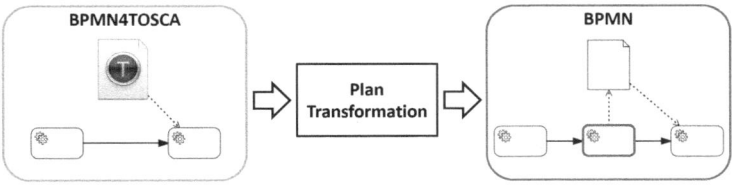

Fig. 3. Injection of Service Tasks (framed in red color) for Accessing Data

are injected. They request information from the TOSCA container and write the information to the data objects as shown in Fig. 3. Vice versa, if BPMN tasks modify data in TOSCA Data Objects, the modified information is sent to the TOSCA container via additional service tasks, too. To achieve this in a coherent way, the TOSCA container offers standardized interfaces to access and modify property information of nodes and relationships.

7 Prototype

Valesca[4] is a modeling tool with full support for TOSCA. Valesca uses the Signavio Core Components[5], which are the commercially-supported enhancements of Oryx [10]. Besides creating service templates, Valesca supports creation of custom node types and relationship types. In the BPMN plan modeling component, the BPMN4TOSCA tasks and data object are added the palette.

When dragging a TOSCA Topology Management Task (cf. Sect. 5.1) from the palette into the plan, the modeling tool suggests a list of topology management operations offered by the container.

When dragging a TOSCA Node Management Task (cf. Sect. 5.2) from the palette into the plan, the modeling tool lists all node templates contained in the topology. After selecting one template, the tool lists all corresponding management operations to let the modeler select the appropriate one. Then, a TOSCA Node Management Task is created having the TOSCA attributes set accordingly. Management operations need input parameters and return values. BPMN foresees the usage of `dataInput` and `dataOutput` [2, pp. 213 and 231]. The data types of the input and output are corresponding to the input and output parameter types defined in the node template's management operation and are generated automatically by the modeling tool.

When dragging a TOSCA Script Task (cf. Sect. 5.3) from the palette into the plan, the modeling tool lists all node templates contained in the topology. After selecting one template, the tool lists all corresponding management operations with script operations as implementation to let the modeler select the appropriate one. The modeler may also choose to discard the choice to specify a script stored directly in the script task.

The modeling tool supports TOSCA Data Objects (cf. Sect. 5.4) in two ways: (i) It provides a separate TOSCA Data Object element in its palette and (ii) manages the corresponding schemas of the properties. When dragging a TOSCA Data Object from the palette into the plan, the modeling tools lists all node templates and relationship templates contained in the topology. One can be chosen to have their property data reflected via the data object. Properties are stored as XML documents defined by an XML schema document (XSD) [1, Sect. 4.2 and 5.2]. In the BPMN modeling tool the TOSCA Data Objects must be typed with the XSD to enable the modeler to extract and process information contained in the

---

[4] http://www.cloudcycle.org/valesca/
[5] http://code.google.com/p/signavio-core-components/

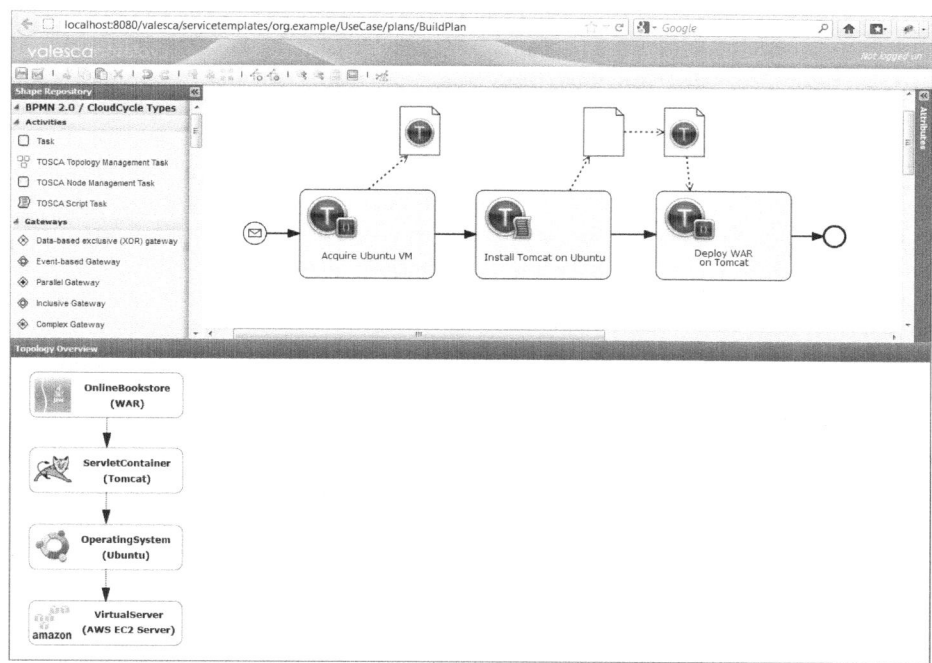

Fig. 4. Deploying the Example Bookstore Application: BPMN4TOSCA Plan in Valesca

properties. Therefore, the modeling tool supports recognizing TOSCA Data Objects, which are syntactically only identified by the additional TOSCA-related attributes, and injects the corresponding `itemDefinitions` whose `structureRef` attributes reference the corresponding XSD.

Besides offering a palette, the modeling tool offers drag'n'drop from the topology model to the BPMN plan model. At the drop of a node template, Valesca asks the modeler to whether he wants to add a TOSCA Data Object or a TOSCA Node Management Task. After the selection, Valesca continues as described in above. The dragging area is shown in Figure 4, which also presents the deployment plan for the use case.

8 Related Work

Extending modeling languages is a common technique to tailor them towards specific needs. For instance, there are at least 62 BPEL extensions including modeling and runtime extensions [9]. The classification in [9] shows that there are a number of design time BPEL extensions which are transformed into plain BPEL. Zor et al. [11] propose a BPMN extension for the manufactoring domain to explicitly handle products and resources. The inclusion of security aspects, such as access control or intrusion detection, into BPMN is described by

Rodríguez et al. [12] through a set of new annotations. However, the presented BPMN extensions do not address the execution of the extended BPMN processes and, therefore, not the transformation to an executable format.

Brucker et al. [13] present SecureBPMN, which is a methodology for secure and compliant business processes covering modeling and runtime. This includes a BPMN extension to add requirements such as access control or separation of duty into the process model. To address the business process runtime, Brucker et al. use a model-based approach to push the security and compliance requirements as configurations into existing systems. The tool chain was prototypically implementation based on Activiti[6], extending the Eclipse designer and process engine.

Discussions on business process transformations usually regard the business-IT-gap and transform high level processes into lower level, more technical, business processes. For instance, Stein et al. [14] survey on transformations to BPEL. Due to the fact that BPMN 2.0 process models are executable, they can be directly deployed to workflow engines for execution. Typically, BPMN 2.0 is transformed to a proprietary meta model of the workflow engine [15].

Before the standardization of TOSCA there have been different approaches in research and practice to model composite applications: Cafe [16] uses a declarative application model to deploy composite application which defines a depends-on and deployed-on relationship. Two approaches using UML [17] to describe the applications or architectures are presented by Machiraju et al. [18] and Arnold et al. [19]. An extensive overview on related work in the field of composite application and enterprise topology modeling is presented by Binz et al. [20].

9   Conclusions and Outlook

In this paper we motivated the upcoming OASIS standard TOSCA and showed how management plans enable the portability of services and their management. We argued that the integration between the service topology and management plans is important for the service creator modeling TOSCA service templates. To offer a tight integration, we propose a BPMN extension called BPMN4TOSCA adding four TOSCA-specific elements to BPMN: (i) The *TOSCA Topology Management Task* to access the TOSCA service topology from management plans, (ii) the *TOSCA Node Management Task* to invoke management operations of nodes, (iii) the *TOSCA Script Task* to execute scripts defined in the TOSCA service topology on nodes, and (iv) the *TOSCA Data Object* to read and write properties of nodes and relationships. We described the integration into a modeling tool and prototypically implemented our approach in the TOSCA modeling tool Valesca. Due to the fact that TOSCA containers use standard BPMN-compliant workflow engines, we showed how to transform BPMN4TOSCA into plain BPMN.

The set of TOSCA topology management operations is not yet fixed. In the context of the CloudCycle project we are working on an open source TOSCA

---

[6] http://www.activiti.org

container, which will provide new management operations to be included in Valesca. The industry partners of the CloudCycle project have started to use Valesca. We are going to use their feedback to further improve user's experience in modeling TOSCA with Valesca.

Acknowledgments. This work was funded by the BMWi project CloudCycle (01MD11023).

References

1. OASIS: Topology and Orchestration Specification for Cloud Applications Version 1.0 Working Draft 07 (June 2012), `https://www.oasis-open.org/committees/download.php/46274/TOSCA-v1.0-wd07.zip`
2. Object Management Group (OMG): Business Process Model and Notation (BPMN) Version 2.0, OMG Document Number: formal/2011-01-03 (2011)
3. OASIS: Web Services Business Process Execution Language Version 2.0 – OASIS Standard (2007)
4. Binz, T., Breiter, G., Leymann, F., Spatzier, T.: Portable Cloud Services Using TOSCA. IEEE Internet Computing 16(03), 80–85 (2012)
5. Kopp, O., Martin, D., Wutke, D., Leymann, F.: The Difference Between Graph-Based and Block-Structured Business Process Modelling Languages. Enterprise Modelling and Information Systems 4(1), 3–13 (2009)
6. Cardoso, J.: How to Measure the Control-flow Complexity of Web Processes and Workflows. In: Workflow Handbook 2005, pp. 199–212 (2005)
7. Reijers, H.A., Mendling, J.: Modularity in Process Models: Review and Effects. In: Dumas, M., Reichert, M., Shan, M.-C. (eds.) BPM 2008. LNCS, vol. 5240, pp. 20–35. Springer, Heidelberg (2008)
8. Weikum, G., Vossen, G.: Transactional Information Systems. Morgan Kaufmann Publishers (2002)
9. Kopp, O., Görlach, K., Karastoyanova, D., Leymann, F., Reiter, M., Schumm, D., Sonntag, M., Strauch, S., Unger, T., Wieland, M., Khalaf, R.: A Classification of BPEL Extensions. Journal of Systems Integration 2(4), 2–28 (2011)
10. Decker, G., Overdick, H., Weske, M.: Oryx – An Open Modeling Platform for the BPM Community. In: Dumas, M., Reichert, M., Shan, M.-C. (eds.) BPM 2008. LNCS, vol. 5240, pp. 382–385. Springer, Heidelberg (2008)
11. Zor, S., Leymann, F., Schumm, D.: A Proposal of BPMN Extensions for the Manufacturing Domain. In: Proceedings of the 44th CIRP Conference on Manufacturing Systems (2011)
12. Rodríguez, A., Fernández-Medina, E., Piattini, M.: A BPMN Extension for the Modeling of Security Requirements in Business Processes. IEICE Transactions on Information and Systems 90(4), 745–752 (2007)
13. Brucker, A.D., Hang, I., Lückemeyer, G., Ruparel, R.: SecureBPMN: Modeling and Enforcing Access Control Requirements in Business Processes. In: ACM Symposium on Access Control Models and Technologies (2012)
14. Stein, S., Kühne, S., Ivanov, K.: Business to IT Transformations Revisited. In: Ardagna, D., Mecella, M., Yang, J. (eds.) BPM 2008 Workshops. LNBIP, vol. 17, pp. 176–187. Springer, Heidelberg (2009)

15. Leymann, F.: BPEL vs. BPMN 2.0: Should You Care? In: Mendling, J., Weidlich, M., Weske, M. (eds.) BPMN 2010. LNBIP, vol. 67, pp. 8–13. Springer, Heidelberg (2010)
16. Mietzner, R.: A Method and Implementation to Define and Provision Variable Composite Applications, and its usage in Cloud Computing. Dissertation, University of Stuttgart, Germany (August 2010)
17. OMG: Unified Modeling Language, UML (2011), http://www.omg.org/spec/UML
18. Machiraju, V., Dekhil, M., Wurster, K., Garg, P., Griss, M., Holland, J.: Towards generic application auto-discovery. In: IEEE/IFIP Network Operations and Management Symposium (2000)
19. Arnold, W., Eilam, T., Kalantar, M., Konstantinou, A.V., Totok, A.A.: Pattern Based SOA Deployment. In: Krämer, B.J., Lin, K.-J., Narasimhan, P. (eds.) ICSOC 2007. LNCS, vol. 4749, pp. 1–12. Springer, Heidelberg (2007)
20. Binz, T., Fehling, C., Leymann, F., Nowak, A., Schumm, D.: Formalizing the Cloud through Enterprise Topology Graphs. In: Proceedings of 2012 IEEE International Conference on Cloud Computing (2012)

All links were last followed on June 29, 2012.

Event-Based Gateways:
Open Questions and Inconsistencies

Felix Kossak, Christa Illibauer, and Verena Geist

Software Competence Center Hagenberg GmbH,
Softwarepark 21, 4232 Hagenberg, Austria
{felix.kossak,christa.illibauer,verena.geist}@scch.at
http://www.scch.at

Abstract. We discuss ambiguities and inconsistencies in the Business Process Model and Notation (BPMN) 2.0 standard regarding the semantics of event-based gateways, and instantiating event-based gateways in particular. We suggest to use instantiating parallel event-based gateways to model asynchronous process behaviour and to clarify the BPMN standard accordingly. We further discourage from any other use of instantiating event-based gateways, and call for a clarification of the semantics of event-based gateways in general.

Keywords: business process modelling, BPMN, event-based gateway, instantiating event-based gateway, semantics.

1   Introduction

The authors have been working on a formalisation of the Business Process Model and Notation (BPMN) 2.0 standard [1] for some time. During this work, despite an intense study of the BPMN standard, we are still uncertain about the intended semantics of some BPMN elements. Often the BPMN standard specifies an element in a very general way in one place, and then constrains this description in various other places. After putting all the different descriptions of one element together, we have sometimes found apparent inconsistencies or even contradictions, while at the same time, the semantics of certain elements remains ambiguous. Studying further literature often confirmed that certain parts of the BPMN standard can be interpreted in different ways, while certain constructs seem to be ignored by the literature and by tools. Sometimes additional literature even added to our confusion.

In this paper, we want to discuss some of the problems we have encountered, namely those concerning event-based gateways, and instantiating event-based gateways in particular. More specifically, we discuss basic issues concerning the activation of events or receive tasks in an event-based gateway configuration, the point of time for creating process instances by instantiating event-based gateways, the instantiation of sub-processes by event-based gateways, and the meaning of "waiting" for instantiating parallel event-based gateways.

J. Mendling and M. Weidlich (Eds.): BPMN 2012, LNBIP 125, pp. 53–67, 2012.
© Springer-Verlag Berlin Heidelberg 2012

After a short overview over related work in Section 2, we thoroughly discuss the issues mentioned above in Section 3 and sum our discussion and proposed solutions up in Section 4.

Please note that section numbers in square brackets, without an additional reference to a certain publication, always refer to the corresponding section in the BPMN standard.

2   Related Work

We did not find publications describing the semantics of all event-based gateway types specified in the BPMN standard [1]. However, amongst others, we found an approach by Dijkman et al. which uses a formal mapping from BPMN to Petri nets in order to be able to statically analyse business process models and check their semantic correctness [2]. The approach deals with a comprehensive subset of BPMN but does not consider instantiating event-based gateways. Cervantes also presents a mapping to Petri nets [3] where a mapping for event-based gateways is included, though not for instantiating event-based gateways. Several other formalisations of BPMN exist which do not regard event-based gateways.

Various different approaches exist for previous versions of the BPMN standard, e.g., a mapping to YAWL [4] or a mapping to BPEL [5]. Weidlich et al. present the other perspective of the latter alignment, i.e. a BPEL-to-BPMN mapping and its pitfalls [6]. For example, they show that the *pick* activity, with the attribute *createInstance* set to "no", can be directly mapped to its counterpart in BPMN, the event-based gateway. However, they indicate compatibility issues concerning the process instantiation mechanisms of BPEL and BPMN, i.e., BPEL scenarios involving multiple start activities are only partially mappable to instantiating event-based-gateways.

Nicolae et al. apply a high-level modelling based on UML to provide a common understanding (in terms of an abstract syntax) of the involved concepts of Service Interaction Patterns, which are directly supported by BPMN 1.1 [7].

Russell et al. propose the *deferred choice* pattern [8] as one of their basic state-based patterns, where the moment of choice, i.e. the decision about which course of action should be taken, is deferred to a later time and based on external factors. (More specifically, there rather is a race between different event triggers or messages than an explicit choice.) This behaviour corresponds to the BPMN exclusive event-based gateway and does not consider issues such as process instantiation or waiting for multiple parallel branches. According to [8], the deferred choice pattern represents a complex pattern that, interestingly, seems to be successfully supported by token-based approaches. E.g., YAWL directly supports this pattern. In UML Activity Diagrams, fork nodes followed by interruptible activity regions can be applied to support deferred choice [9].

In [10] the *racing incoming messages* pattern is described using exclusive event-based gateways. Furthermore, the deferred choice pattern is extended by reaction rules attached to additional rule gateways in order to be able to constrain the decisions made by the environment. In doing so, the activation of activities can dynamically be determined using predefined conditions.

Another approach uses the event calculus to discuss a pattern for the event-based split, where the decision is based on the occurrence of events as in the case of the BPMN exclusive event-based gateway [11].

Hofstede et al. also consider the exclusive event-based gateway to be a useful pattern for process integration, where activities should only be activated if certain preconditions triggered by events are met (*event-based task enablement* pattern) [12]. In addition, they identify the need for a mechanism to instantiate a process on the receipt of an event (*event-based process instance creation* pattern).

The introduction of "event-based gateways as start nodes" in BPMN was proposed by Decker and Mendling [13] in order to cover the "Reachable Subscription" for process instantiation according to their CASU framework for describing instantiation designs.

Similar to our work, Börger and Sörensen [14] formalise the semantics of BPMN (Version FTF Beta 1 for Version 2.0 [15]) using Abstract State Machines (ASMs) [16], where the formalisation is kept relatively abstract. For the event-based gateway, Börger and Sörensen use a group concept that is not defined in the BPMN 2.0 standard.

3   Particular Issues Regarding Event-Based Gateways

In this section we discuss four basic issues concerning event-based gateways, with a focus on instantiating event-based gateways. In particular, we discuss the activation of events or receive tasks in an event-based gateway configuration, the point of time for creating process instances by instantiating event-based gateways, the instantiation of sub-processes by event-based gateways, and the meaning of "waiting" for instantiating parallel event-based gateways.

3.1   Activation of Events or Receive Tasks

The first question we want to raise is: When should an event-based gateway send tokens to the events or to the receive tasks in its configuration? While this may appear to be a minor technical issue, any answer to this question will have repercussions on the question of the timing of process instantiation as discussed in Section 3.2.

We start the discussion with *non*-instantiating gateways, which is the simpler case. Note that non-instantiating gateways can only be exclusive gateways. Consider the situation when a token has reached the event-based gateway.

The BPMN standard says that "The choice of the branch to be taken is deferred until one of the subsequent **Tasks** or **Events** completes" [13.3.4]. This appears to suggest that *no* tokens are sent by the gateway to any of the events or receive tasks in its configuration. However, it is stated elsewhere in the standard that a receive task needs to be activated before it can start waiting for a message [13.2.3]; but also an intermediate event (the only event type possible here) is obviously supposed to get a token before it can start waiting for events: "Waiting starts when the **Intermediate Event** is reached" [13.4.2].

The question whether an event-based gateway should pass tokens on *immediately* or only *after a path has been chosen* will be further complicated in the case of instantiating event-based gateways (see further below), and we will have to make sure that any solution is consistent for both instantiating and non-instantiating gateways. The question does have practical consequences for implementations, which become obvious when we try to translate the specification into algorithms. But problems for implementation can as well hint at possible uncertainties when humans try to understand the constructs in question.

According to the first interpretation, in which tokens are passed on immediately, the sequence of actions will be as follows:

- On receiving a token, the gateway sends tokens towards all events or receive tasks in its configuration (like a parallel gateway).
- When the first event receives a trigger or the first receive task receives a message, this node passes its token on, while other tokens on sequence flows to the other events are deleted and other receive tasks are interrupted (i.e. their task instances are "withdrawn", cf. Fig. 13.2 of the standard).

Note that deleting a token from a sequence flow without passing it on is usually only done by end events. Thus, either the intermediate events that were not triggered would have to behave like end events in such a case, or the respective tokens would have to be withdrawn by the gateway. We think that both solutions represent an extraordinary behaviour, thus impairing intelligibility.

In the case of receive tasks, the solution becomes clear when we look at the lifecycle model for activities [13.2.2]: the receive task is interrupted and the lifecycle state is set to "Withdrawn". This explicit provision in the lifecycle model (see Fig. 13.2 in the standard) is a strong hint that the intended solution (at least according to [13.2.2]) is to forward tokens *immediately* and subsequently to withdraw all but one of them.

However, besides the unusual behaviour of events, the solution has the disadvantage that control is taken out of the hands of the gateway itself. Should two or more events receive a trigger each at the same time, both would fire, leading to two or more different paths taken by the same process instance – something which is clearly not intended, as the gateway should behave like an *exclusive* gateway [13.3.4]. (In the case of receive tasks, two or more messages could arrive at the same time, basically leading to the same dilemma.)

A solution to this could be to defer firing of the event (despite having received a token *and* an event trigger) until the gateway responds with an "okay" (the gateway might make a choice in the case of concurrent triggers to avoid *subsequent* concurrency). But surely the intuition of passing on a token is that of passing on control, and breaking with such an important intuition would impair the intelligibility of BPMN process diagrams in general.

The other basic choice would be for the gateway to defer passing on tokens to events and receive tasks in its configuration until the first event trigger or message has been registered. In this case, *the gateway itself* would have to wait for event triggers and/or messages, leading to the following overall behaviour:

- Wait for one of a set of event triggers or for a message for a receive task in the configuration of the gateway; if applicable, also check correlation information.
- Once such a trigger or message has occurred, select the respective path.
- Send a token down the selected path, *and* resend the trigger or message (as both events and receive tasks can only start waiting once they have received a token).

The solution sketched above has the following drawbacks:

- The gateway must have all the "intelligence" to filter applicable event triggers and, if required, to process correlation information.
- The lifecycle state of "Withdrawn", all paths to which being labelled with "An Alternative Path for Event Gateway Selected" (see Fig. 13.2 in [13.2.2]), would become superfluous.
- Resending the event trigger or message constitutes an extraordinary behaviour (as sending a message within the same process is a breach of the standard).

A third option is to treat the event-based gateway plus the events and receive tasks in its configuration as *one single node*. The respective event nodes and receive tasks would then be relegated to graphical symbols with no separate semantics. This would do away with having to resend the trigger or message, but the two other drawbacks noted above would still hold.

Summarising the three options discussed above, we note that there are strong arguments for the first option – especially the lifecycle model presented in [13.2.2] as well as the fact that both intermediate events and receive tasks usually need tokens to start waiting. However, we want to defer a concluding judgement for now. Having these three options in mind, with their advantages and drawbacks as identified so far, we now turn to *instantiating* exclusive event-based gateways.

The relevant difference is that the gateway itself does not get any token to pass on, for no process instance is existing as yet before a respective event trigger or message has been received. So if we still take the first option discussed above into account, the gateway would already have to send tokens to events and receive tasks in its configuration *before any process instance had been triggered* from outside.

Certainly, this would be feasible for a process engine: whenever the engine is started, all instantiating event-based gateways are already activated (i.e. a process instance is created), and whenever an instantiating event-based gateway actually fires, it is immediately re-activated with a new process instance. However, we think that this would constitute a very extraordinary behaviour, and we do not believe that such a behaviour was intended. Note that there will always be tokens on some sequence flows or activated receive tasks whose instance will never complete.

So what seemed to be the most plausible alternative for interpreting the BPMN standard with respect to the exact behaviour of event-based gateways now seems implausible. As we do not want to make principal differences between non-instantiating and instantiating event-based gateways in this respect (which would break intuition), we are effectively left with three alternatives for interpreting the BPMN standard, each of which seems implausible for different reasons. However, the third option – to treat event-based gateways plus the events and receive tasks in its configuration as a single node – now seems to be the best solution, requiring that the lifecycle state of "Withdrawn" be removed from Fig. 13.2 in [13.2.2].

There is another type of event-based gateways, which is the instantiating *parallel* event-based gateway. Here we do not have to withdraw tokens or receive task instances, which would make implementation of the first variant (sending out tokens immediately) easier. However, the last problem discussed for instantiating event-based gateways remains, namely that tokens would have to be sent before any process instance was triggered. Thus, this case does not change the situation or give us significant, additional insight. (We will not go into the issue of correlation – which is required for instantiating parallel gateways – in this place, because it does not appear to be relevant for our problem.)

3.2   When Is a Process Instantiated by an Instantiating Event-Based Gateway?

The above discussed problem is closely related to the question of when exactly a process is instantiated by event-based gateways, to which we turn now.

We revert to our previous observation that both intermediate events and receive tasks obviously need to be "activated", i.e. have received a token (with the obvious exception of instantiating receive tasks) before they can wait for event triggers or messages. (Note that for instantiating receive tasks, the problem of activation is the same as for receive tasks in the configuration of instantiating event-based gateways.) This would suggest that, if we rule out the option to create tokens before an instance is triggered (see Section 3.1), the following actions would have to be taken by an instantiating gateway upon the occurrence of a suitable event trigger or message:

- In case of an event trigger, check the trigger type; if it fits, do the following, else skip. In the case of a message, do the following.
- Activate the event-based gateway (and thereby create a process instance) and send a token to the respective event or receive task (or, in the "one single node" alternative, directly towards a node following the latter).
- Forward (or re-send) the trigger or message to the respective node (not necessary in the "one single node" alternative, see Section 3.1).

This is basically the same algorithmic sketch as that for the second alternative in Section 3.1, only adapted to stress the point of activation. The instantiating event-based gateway is activated *after* a trigger or message has been sent, and thereby, at the same time, a new process instance is created. The drawbacks of this interpretation have already been discussed.

An alternative would be to perform the following actions *by the agent that sends* the respective event trigger or message:

– First, activate the appropriate instantiating event-based gateway of the target process (i.e. create a process instance).
– Only then send the event trigger or message.

However, this would require a sending agent to know about the internal structure of the target process. We think that this is an obstacle to modularised development of business processes. We argue that this is violating the principle of information hiding, as together with the gateway, we also have to reveal the branching behaviour, which should be regarded as a matter of implementation rather than a part of the interface. In any case, this *hidden* behaviour of instantiating a process before sending an event trigger or message to it seems highly unintuitive.

But this solution gives a hint, in our opinion, that it would be a better idea for process designers to use a start event in front of an event-based gateway instead of using an *instantiating* (exclusive) event-based gateway, for this would re-establish the intuition of instantiating a process *explicitly* before communicating with it. (Note that the same algorithmic sketch, as well as the problem involved with it, would apply to instantiating receive tasks as well.) However, the situation is different for instantiating *parallel* event-based gateways, which we discuss in Section 3.4.

Summing up, the question as to when exactly a process is instantiated by an instantiating event-based gateway is closely tied to the question of its internal behaviour as discussed in Section 3.1. The drawbacks of the "constant instantiation" solution are basically the same, as are the drawbacks of the "instantiate after receipt" solution. The extra alternative of *external* (hidden) instantiation exhibits its own problems.

The use of start events constitutes an obvious solution – at least in the case of exclusive gateways (but see Section 3.4) – and would also mitigate the problems discussed in Section 3.1, because the first option discussed there (to send tokens to event nodes and receive tasks in the configuration of the gateway immediately) still seems plausible and viable for *non-instantiating* event-based gateways. Consequently, we suggest to *not use instantiating exclusive event-based gateways* – and strike them off the BPMN standard.

3.3   Starting Sub-processes Using Event-Based Gateways

May an instantiating event-based gateway instantiate a sub-process? In the context of sub-process instantiation, the standard states:

> A **Sub-Process** is instantiated when it is reached by a **Sequence Flow** *token*. **The Sub-Process** has either a unique empty **Start Event**, which gets a *token* upon instantiation, or it has no **Start Event** but **Activities** and **Gateways** without *incoming* **Sequence Flows**. In the latter case all such **Activities** and **Gateways** get a *token*. A **Sub-Process** MUST not have any non-empty **Start Events**. [13.2.4]

This would allow an instantiating event-based gateway to start a sub-process, because it is a gateway without incoming sequence flows. The only constraint is that the sub-process must not have a start event in addition to the instantiating event-based gateway.

Elsewhere, the standard states that (instantiating) event-based gateways constitute "the only scenario where a **Gateway** can exist without *incoming* **Sequence Flows**" [13.4.1]. Consequently, this is the only type of gateway which could possibly satisfy the condition for [13.2.4] (see above).

Due to the fact that the first arrival of an event trigger or a message for the gateway creates the instance of the sub-process (but see also the discussion in Section 3.2), one could interpret such a trigger or message as the trigger of an implicit start event; however, an implicit start event is not allowed to have a trigger by the BPMN standard:

> If a **Start Event** is not used, then the implicit **Start Event** for the
> **Process** SHALL NOT have a *trigger*. [10.4.2]

However, it is not really clear if an event-based gateway may start a sub-process, because all other provisions of the BPMN standard concerning event-based gateways only mention "starting a process":

> A **Process** can also be started via an **Event-Based Gateway**
> or a **Receive Task** that has no *incoming* **Sequence Flows** and its
> instantiate flag set to *true*. [13.1]

> A Process can also be started via an Event-Based Gateway, [...]
> [10.4.6]

A further indication that a sub-process cannot actually be started by an event-based gateway is that not a token, but the first arrival of an event trigger or a message would create a process instance, which contradicts [13.2.4] (see above, first sentence); but then why are gateways mentioned for starting sub-processes?

The only possible solution to implement the provisions of [13.2.4] with respect to event-based gateways is that first, the sub-process is instantiated upon receiving a token, whereupon (in the absence of a start event) its event-based gateways without incoming sequence flows each get a token. But then the event-based gateways would not be instantiating any more, and consequently would need to have incoming sequence flows. (Note that this solution would also suggest the "external instantiation" option discussed but advised against by us in Section 3.2.)

One solution, though only for instantiating *exclusive* event-based gateways, would be to add a start event in front of the event-based gateway, as already suggested in Section 3.2. This would leave the case of instantiating *parallel* event-based gateways, because these are always instantiating and can never have incoming sequence flows:

> The **Event Gateway**'s instantiate attribute MUST be set to *true*
> in order for the eventGatewayType attribute to be set to Parallel [...]
> [10.5.6]

Another solution, but again only for instantiating *exclusive* event-based gateways, would be to use several start events instead of an event-based gateway, as demonstrated by Allweyer [17]. However, Allweyer refers to the fact that several start events can lead to misinterpretations, which might be avoided using instantiating event-based gateways. Allweyer states that both models in Fig. 1 and Fig. 2 have the same semantics. But several start events can only be used for top-level processes (see [13.2.4], which requires a "unique" start event for sub-processes; also note that [13.2.4] also states that this start event must be "empty").

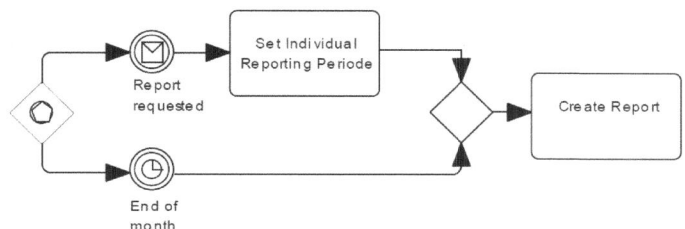

Fig. 1. Instantiating exclusive event-based gateway (source: [17] (Fig. 101))

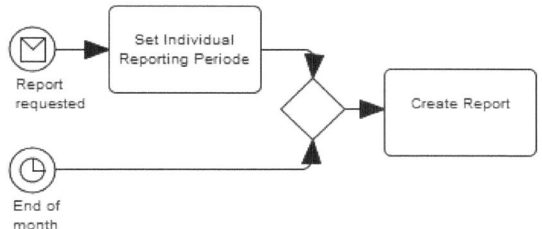

Fig. 2. Several alternative start events (source: [17] (Fig. 73))

A solution for handling instantiating parallel event-based gateways in sub-processes would be to use a start event with a "None" trigger followed by a parallel gateway (instead of a parallel event-based gateway) where each path waits for an event and then does some work (see Fig. 3). A parallel merge gateway is inserted before the end event for synchronisation, but this is not really necessary, because the tokens produced at the parallel split gateway would avoid leaving the sub-process as long as not all paths are executed.

The only difference to using an instantiating parallel event-based gateway is that the sub-process is instantiated when it is reached by a token and not when the first event trigger or message arrives. However, a sub-process must be activated by a token anyway (see [13.2.4]), except for an Event Sub-process, which "MUST have one and only one **Start Event**" [10.2.5] (note, however,

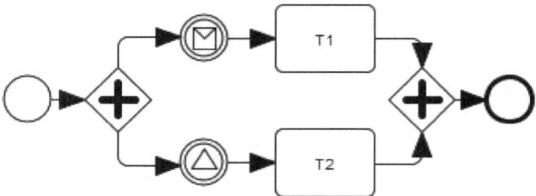

Fig. 3. Sub-process with a behaviour similar to an instantiating parallel event-based gateway

that this is not reflected in [13.2]). This appears to make instantiating event-based gateways redundant for sub-processes. Accordingly, we suggest to remove the possibility of sub-processes having "**Gateways** without *incoming* **Sequence Flows**" from [13.2.4].

3.4   On the Meaning of Waiting for Process Instances Started by Parallel Event-Based Gateways

In certain cases a process should be instantiated when the first message of a set of messages required for the work of a process instance is received. The BPMN standard, therefore, introduces a further variant of the event-based gateway, i.e. the instantiating parallel event-based gateway. This gateway "MUST not have any *incoming* **Sequence Flows**", the gateway's "instantiate attribute is set to *true*", and the "eventGatewayType attribute is set to Parallel" [10.5.6].

A process which defines a parallel event-based gateway as a process start is instantiated when the first event in the gateway configuration is triggered or the first message for a receive task arrives, respectively; however, the other events or receive tasks in the gateway configuration remain enabled. In this case, events are only allowed to have message-based triggers which must share exactly the same correlation information [10.5.6]. After the process has been instantiated upon the occurrence of the first message trigger, the remaining triggers will belong to the already created process instance, i.e. no new process instances will be created [13.3.4]. In the BPMN standard, the behaviour of a process that was created by the first matching event trigger of an instantiating parallel event-based gateway is described as "the **Process** then waits for the other **Events** to arrive" [13.1].

However, when specifying the semantics of the instantiating parallel event-based gateway in our work, the question of what exactly *waiting* should mean arose. Does *waiting* mean that the node where the first event trigger or message occurred receives a token on each of its outgoing sequence flows, and the other events or receive tasks get a token on their incoming sequence flows (in order to cause waiting)? This assumption seems to be feasible, because the event-based gateway is a splitting and not a merging gateway. It might further be corroborated by another statement of the BPMN standard: "the other **Events** are still waiting and are expected to be triggered before the **Process** can (normally) complete" [10.5.6]. Or should the term *waiting* be interpreted so that all outgoing sequence flows of all events (including that where the first event trigger occurred) get a token only after all events have occurred? The already stated definition, "the **Process** then waits for the other **Events** to arrive" [13.1], also tolerates this interpretation.

According to [18], using an instantiating parallel event-based gateway expresses that all events or receive tasks in the gateway configuration must have occurred or completed, respectively, before a process can be started *completely*. For example, a broker independently receives buying orders and offers for sale. However, both of them need to be available to successfully process the sale. The instantiating parallel event-based gateway also handles correlation, which cannot be achieved using multiple start events with a simple AND join. Thus, the two models shown in Fig. 4 and Fig. 5 cause different behaviours. In the first model, the occurrence of the first event instantiates the process, and the created token is then merged with the token created upon the occurrence of the second event within the same process instance (correlation). In the second model, however, two isolated process instances would be created, each creating a token that would wait forever at the AND join. (Note that multiple alternative start events can be merged just as well by an XOR join without the need for using an instantiating exclusive event-based gateway.) Looking at the example given, upon the arrival of a buying order, the gateway checks for an existing process instance that was instantiated by a fitting offer for sale and vice versa.

Freund et al. [18] also suggest a further approach to start a process by two or more events (which is also possible for BPMN 1.2). Thereby, all distinct combinations concerning the ordering of events need to be modelled separately, e.g., a buying order as well as an offer for sale may instantiate the process,

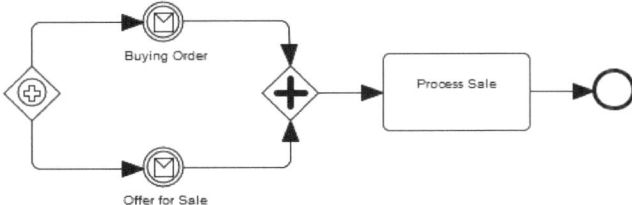

Fig. 4. Parallel event-based gateway to start a process (source: [18], slightly modified)

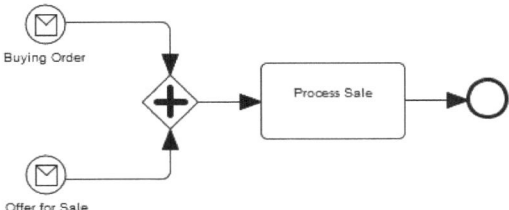

Fig. 5. Multiple, parallel start events to start a process

and in each case correlation can be handled (see Fig. 6). However, why does the instantiating parallel event-based gateway exist anyway? Freund et al. [18] argue that some studies revealed that process models with exactly one start node are more intelligible. Moreover, according to [17], process models with multiple start events can easily lead to misinterpretations (see also Section 3.3). Moreover, with more than two messages to wait for, the diagram will become considerably cluttered with redundant elements.

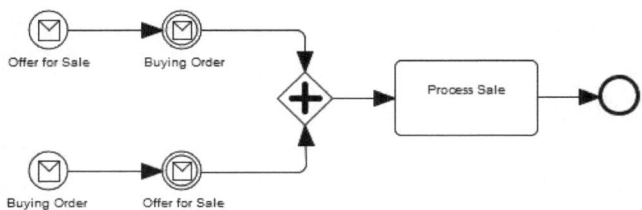

Fig. 6. Combinations of distinct events to start a process

Considering the notion of *waiting* for instantiating parallel event-based gateways, Allweyer states in [17] that "the process waits for the other events before it is continued" and "all events after the gateway must occur at the beginning". He also states that the behaviour of the instantiating parallel event-based gateway can be modelled with a parallel multiple start event as well. In conformity with the BPMN standard, for a parallel multiple start event "there are multiple *triggers* REQUIRED before the **Process** can be instantiated" [10.4.2]. Regarding our broker example, this means that as long as only a buying order is received, it is necessary to wait for the arrival of the offer for sale and vice versa before the sale can be processed. Thus, the model shown in Fig. 7 is equivalent to the models shown in Fig. 4 and Fig. 6 respectively.

Fig. 7. BPMN parallel multiple start event to start a process

Again, this raises the question why a new element is introduced in BPMN 2.0 that (nearly) exhibits the same behaviour as existing ones. (Note that there is a slight difference concerning the point in time when the process is instantiated, i.e., using a parallel multiple start event causes the process to be instantiated after all listed types of triggers have occurred, whereas an instantiating parallel event-based gateway instantiates the process when the first event trigger or message is received. However, this reaction to different events already describes a detailed course of operations that is not relevant for the basic process flow.) Regarding the use of instantiating parallel event-based gateways in the literature, this newly introduced element seems to be replaceable by existing BPMN elements in case that a synchronisation takes place right after the arrival of all required event triggers or messages. However, we expect the instantiating parallel event-based gateway to be more than syntactic sugar, which is the case if the synchronisation of enabled, independent paths is realised at a later point in time within the process. This way, the instantiating parallel event-based gateway would provide a simple and subtle way to perform processes asynchronously. We consider such an application as a major advantage in business process modelling.

4 Summary

In this paper, we have documented several issues concerning event-based gateways as described in the BPMN standard where the exact semantics seems ambiguous and sometimes even contradictory.

We have drawn attention to problems with triggering event-based gateways (Section 3.1) and, closely related, with determining the moment when an event-based gateway should be considered to have been triggered and, in the case of an instantiating gateway, when a respective process instance should be created (Section 3.2). All the interpretations of the BPMN standard which we could think of turned out to contradict some part of the standard.

Further problems concern the use of event-based gateways to instantiate sub-processes, where we discovered further discrepancies between different parts of the BPMN standard (Section 3.3).

While the general idea of event-based gateways seems intuitive at first sight, the semantic details are far less so, most of all in the case of instantiating event-based gateways. Clearly, the standard needs to clarify many points regarding the semantics of event-based gateways, and especially sections 10 and 13 of the standard need to be rendered consistent with each other.

Also the interpretation of the semantics of instantiation by parallel event-based gateways is ambiguous in a crucial way, namely concerning the question whether it allows for asynchronous behaviour of different branches after the gateway or not (Section 3.4). This question also affects the question whether such a gateway could be replaced by other constructs (including start events) or not. We think that such instantiating gateways could be useful for modelling asynchronous behaviour of different branches of a process, in which case they obviously cannot be properly replaced by other constructs. However, we find them dispensable if their semantics includes synchronisation. We discourage from any other use of *instantiating* gateways. (Note that the suggestion by Decker and Mendling in [13] to introduce instantiating event-based gateways in BPMN does *not* apply to *exclusive* event-based gateways, as the problem of necessary "subscriptions" – or instance correlation – cannot occur there.)

To finally sum our suggestions up, we

- suggest to use instantiating parallel event-based gateways to model *asynchronous* process behaviour;
- call for a respective clarification of the BPMN standard;
- discourage from any other use of instantiating event-based gateways;
- suggest to drop the use of "gateways without incoming sequence flows" for starting sub-processes (in [13.2.4]) as irrelevant;
- suggest to interpret the semantics of event-based gateways such that tokens are sent to nodes in its configuration *immediately* upon activation of the gateway, and clarify the standard accordingly, except
  - *if* instantiating exclusive event-based gateways *are* retained, then interpret the semantics of event-based gateways such that the gateway and the nodes in its configuration are seen as *one single flow node* instead, which would make it possible to
  - instantiate a process *internally, after* receiving an event trigger or message.

Acknowledgement. This publication has been written within the project "Vertical Model Integration". The project Vertical Model Integration is supported within the program "Regionale Wettbewerbsfähigkeit OÖ 2007-2013" by the European Fund for Regional Development as well as the State of Upper Austria.

This work was also supported in part by the Austrian Science Fund (FWF) under grant no. TRP 223-N23.

References

1. OMG: Business process model and notation (BPMN) 2.0 (2011), http://www.omg.org/spec/BPMN/2.0 (accessed August 2011)
2. Dijkman, R.M., Dumas, M., Ouyang, C.: Semantics and analysis of business process models in BPMN. Inf. Softw. Technol. 50(12), 1281–1294 (2008)
3. Cervantes, A.A.: Representation of the behavior of business process models. Technical report, Faculty of Mathematics and Computer Science University of Tartu, East Lansing, Michigan (2011)
4. Ye, J., Sun, S., Song, W., Wen, L.: Formal semantics of BPMN process models using YAWL. In: 2008 Second International Symposium on Intelligent Information Technology Application, IITA 2008, vol. 2, pp. 70–74. IEEE Computer Society, Washington DC (2008)
5. Ouyang, C., Dumas, M., van der Aalst, W.M.P., ter Hofstede, A.H.M., Mendling, J.: From business process models to process-oriented software systems. ACM Trans. Softw. Eng. Methodol. 19(1), 2:1–2:37 (2009)

6. Weidlich, M., Decker, G., Großkopf, A., Weske, M.: BPEL to BPMN: The Myth of a Straight-Forward Mapping. In: Meersman, R., Tari, Z. (eds.) OTM 2008, Part I. LNCS, vol. 5331, pp. 265–282. Springer, Heidelberg (2008)
7. Nicolae, O., Cosulschi, M., Giurca, A., Wagner, G.: Towards a BPMN Semantics Using UML Models. In: Ardagna, D., Mecella, M., Yang, J. (eds.) BPM 2008 Workshops. LNBIP, vol. 17, pp. 585–596. Springer, Heidelberg (2009)
8. Russell, N., ter Hofstede, A.H.M., van der Aalst, W.M.P., Mulyar, N.: Workflow Control-Flow Patterns: A Revised View. Technical report, BPMcenter.org (2006)
9. Wohed, P., van der Aalst, W.M.P., Dumas, M., ter Hofstede, A.H.M., Russell, N.: Pattern-Based Analysis of the Control-Flow Perspective of UML Activity Diagrams. In: Delcambre, L.M.L., Kop, C., Mayr, H.C., Mylopoulos, J., Pastor, Ó. (eds.) ER 2005. LNCS, vol. 3716, pp. 63–78. Springer, Heidelberg (2005)
10. Milanović, M.V.: Modeling Rule-driven Service Oriented Architectures. Thesis, University of Belgrade (2010)
11. Zahoor, E.: Gouvernance de service: aspects sécurité et données. Thesis, Université Nancy II (2011)
12. ter Hofstede, A.M., van der Aalst, W.M.P., Adamns, M., Russell, N. (eds.): Modern Business Process Automation: YAWL and its Support Environment. Springer, Heidelberg (2010)
13. Decker, G., Mendling, J.: Process instatiation. Data & Knowledge Engineering 68(9), 777–792 (2009)
14. Börger, E., Sörensen, O.: BPMN core modeling concepts: Inheritance-based execution semantics. In: Embley, D., Thalheim, B. (eds.) Handbook of Conceptual Modeling: Theory, Practice and Research Challenges. Springer, Heidelberg (2011)
15. OMG: Business process model and notation (BPMN) FTF beta 1 for version 2.0 (2009), http://www.omg.org/spec/BPMN/2.0/Beta1/ (accessed June 2012)
16. Börger, E., Stärk, R.: Abstract State Machines - A Method for High-Level System Design and Analysis. Springer, Heidelberg (2003)
17. Allweyer, T.: BPMN 2.0. Books on Demand, Norderstedt (2010)
18. Freund, J., Rücker, B., Henninger, T.: Praxishandbuch BPMN: Incl. BPMN 2.0. Hanser, München (2010)

Extending BPMN 2.0
for Modelling the Combination
of Activities That Involve Data Constraints

Luisa Parody, María Teresa Gómez-López, and Rafael M. Gasca

University of Seville, Computer Languages and Systems Department,
Avda. Reina Mercedes s/n, 41012 Seville, Spain
{lparody,maytegomez,gasca}@us.es
http://www.lsi.us.es/~quivir/

Abstract. The combination of activities to achieve optimal goals sometimes has a complex solution. Business Process Model and Notation
(BPMN) 2.0 facilitates the modelling of business processes by providing
new artifacts, such as various types of tasks, source of data and relations
between tasks. Sometimes, although the order of the activities can be
known, the concrete data values that the activities interchange to optimize their behaviour needs to be found, specially when input parameters
of an activity affect to the input parameter of the others. Taking into account the lack of priority and clear sequential relationship between the
activities of such combination, a deep analysis of possible models and
data input values for the activities is necessary. For that reason, an extension of BPMN 2.0 with a new type of sub-process and its associated
marker is proposed. The aim of this new sub-process is to define, in an
easy way, a combination of several activities to find out, in an automated
way, the concrete values of the data handling that optimize an overall
objective.

Keywords: Business Process Management, Business Process Model and
Notation, Combination of Activities, Data Input, Data Constraint.

1 Introduction

A business process, henceforth referred as BP, consists of a set of activities that
are performed in coordination in an organizational and technical environment.
These activities jointly perform a business goal [20]. The combination of activities is very important from the point of view of BP management, since it can
increase customer satisfaction, reducing business investment and establishing
new products and services at low cost.

Generally, a combination between various activities or business processes is
almost always complicated, since every stakeholder focuses on their own inter-
ests. The degree of complexity becomes even greater when the objective of the
combination is to optimize a common business goal. The ideal scenario arises
when all the activities can obtain an optimal result based on the decisions made

J. Mendling and M. Weidlich (Eds.): BPMN 2012, LNBIP 125, pp. 68–82, 2012.

by each activity in an independent manner. The problem comes when values used as data input of an activity depend on the values taken for other activities in the process. Moreover, when the data input given by the user have interval domains (range of possible values), then obtaining the best local result does not necessarily imply that the best global result follows. In that case, how to combine the activities/sub-processes? This problem is worst when there is no priority and clear sequential relationship between the activities, since the artifacts in BPMN 2.0 required a decision order at design time (even in the ad-hoc sub-process the performer has to decide an activity order). Furthermore, the construction of models, in terms of activities order, implies an off-line analysis where the different combinations are studied. However, this off-line analysis is not enough when the behaviour and functionality of the activities are unknown, since they are external services which relationships between data input and output are unknown. For example, to known the price of a flight for a date and a pair of cities, it is necessary to call the service, then an off-line analysis is not enough to find out the best data input for the service. In that case, the relations between data inputs and outputs of activities have to be discovered, thus an analysis of all the possible data combinations and the behaviour of these activities in run-time is needed. In addition, the data combinations must take into account a set of existing data constraints that relate them and an objective function to be optimized. Therefore, the aim of the problem is to decide how to model with BPMN a BP whose activities (i) must be executed many times to search the concrete values for their data input that optimize the overall objective, and (ii) satisfy all the constraints that relate the values of the data input and output. This implies that the model has to be adaptive and flexible to manage the data handling in each instance.

The main standard used to model BP is Business Process Model and Notation (BPMN), proposed by OMG. The new version 2.0 solves the majority of the modelling problems. However, it remains as yet insufficiently powerful since, among other requirements, there is a significant need for the representation of the combination of activities where, the concrete data of each activity depends on the concrete data of the others, since they are related by means of constraints. To the best of our knowledge, there are no solutions with BPMN that enable to represent problems where the explicit order of the activities are not defined, and where the model is influenced by data constraints. Although it represents business processes, by means of Sequence Flows and Data Flows, the objective function and the constraints that relate the data input can only be added through annotations. Moreover, the annotations cannot be mapped to executed code in an automatic way.

For this reason, the aim of this paper is to extend the expressiveness of BPMN 2.0 with a new sub-process. The proposal is focused on the support of this combination of activities, whose main purpose is to optimize an objective function, where there is a set of possible data input for the activities that are related between them by means of constraints. Furthermore, this new sub-process enables the creation of executable code in an automatic way.

The rest of the paper is organized as follows: Section 2 motivates and explains the necessity of a combination of activities to search the concrete values of data to achieve a common goal. Section 3 summarizes the main issues about the combination of activities specification. Section 4 defines the BPMN extension to represent the combination of activities from the business descriptions perspective. Section 5 includes certain relevant related work. And finally, conclusions are drawn and future work proposed in Section 6.

2   Motivating Example

In this section we present a scenario to illustrate the combination of activities in the kind of problems addressed in this paper. Later on, it will be also used to explain our approach.

Our scenario is the Travel Booking Example presented in [13]. This example is proposed to show in-line event handling via event sub-process constructs. Although this is not the purpose of this work, the example highlights the lack of graphic representation for the combination of activities to search the concrete values for the data handling that achieve a common objective.

If we look through the Travel Booking Diagram (Chapter 9, page 28 in [13]), the customer wants to book a flight and a hotel room. The resulting business process (i) starts with the travel booking reservation request, (ii) follows with the search and evaluation of both flights' and hotel rooms' availability, (iii) selected alternatives are packaged and offered to the customer and (iv) the customer can select a proposed alternative or cancel the request. We are focused on the second step, where the searches of flights and hotel rooms are based on the customer request and are made in a parallel way. This is only possible when the customer request is formed by a concrete data input, in other words, when each data input has only one possible value (atomic value) and there is no relation between the customer criteria and the final result of both activities. It means that there are no relations between the activities, then they can be executed in a parallel way. For example, the customer wants to organize a trip, where he wants to depart on 2012-09-15 and return on 2012-09-30, with a flight from Madrid to London, and wants to book a hotel room in London during these days. The dates and the locations given by the customer are the data input. These data input have atomic values for the flight and hotel room searches, hence both searches can be performed in an independent way. Generally, the customer searches a trip for a concrete dates and location. Then, if necessary and also cheaper, the customer tries with another departure or return dates, thereby expanding the range of dates when leaves or arrives.

For that reason, we focus on problems when the customer wants to obtain the best result with a specific criteria by providing constraints between the data given. In that case, executing the activities in a parallel way is not enough, since obtaining the optimized value in an independent manner does not necessarily mean that the overall result will be the best option. For example, the customer wants the cheapest trip with the data input presented before, in this case also providing an interval (either range) of dates (for example, the departure date can be between 2012-09-15 and 2012-09-17, and the return can be either on 2012-09-30 or on 2012-10-01). Both, the flight and hotel room searches, have to combine their results and decisions to obtain the common objective. These possible values are also determined by a set of constraints that relates these data input, for example, (i) the depart date has to be a value between 2012-09-15 and 2012-09-17, (ii) the return date has to be a value between 2012-09-30 and 2012-10-01, and (iii) the return date of the flight and the check-out date at the hotel have to be the same. Therefore, this is an example where the activities share their data input. On the other hand, the objective function is a combination of the data outputs of these activities, given by their executions. In this example, the sum of the cost of buying an airline ticket and stay in a hotel room for these dates and locations. The aim of the problem is to know which is the best set of values for the interval data input that optimize the objective function and satisfy all the constraints.

3   Formal Definitions

In order to clarify the problem, a formalization is introduced. With the aim to understand how the relation between the activities data can be defined, the constrains definition that involve them and the objective function are essential. This definition enables the identification, description and definition of the type of problems that need the modelling for the combination of activities.

Let a **Combination of Activities that involves Data Constraint to Optimize an Objective Function (CAO)** be a business process whose activities $(A_1, ..., A_n)$ are independent and there is no priority order and clear sequence relationship between them. For each activity A_i, a set of input variables (I_{A_i}), and a set of output variables (O_{A_i}) are defined. Then, the following concepts are introduced for this CAO:

Definition 1. Activities Data Input (ADI): *The set of variables that represents the union set of all the data input of the* n *activities of the* CAO.

$$ADI = \bigcup_{i:1..n} I_{A_i}$$

Definition 2. Activities Data Output (ADO): *The set of variables that represents the execution result of the* n *activities* (O_{A_i}).

$$ADO = \bigcup_{i:1..n} O_{A_i}$$

Definition 3. Process Data Input (PDI): *The set of variables that represents the data introduced to the BP. Every variable* $x_j \in PDI$ *could have multiple values* $v(x_j) \in D(x_j)$, *where* $D(x_j)$ *is the domain of* x_j $(D(x_j)$ *is a finite set comprising all possible values that can be assigned to variable* x_j).

Definition 4. *Process Data Output (PDO): The set of variables returned by the BP. These variables represent the concrete values for the ADI that optimize the objective function. These concrete values will be provided to the user of the BP or another external process.*

Definition 5. *Objective Function (ObjFun): The global optimization function to be satisfied. This function can be defined in terms of ADO, ADI and PDI, and can be maximizing or minimizing.*

In a *CAO*, there is a set of **constraints (C)**, where each $C_k \in C$ relates a subset of variables $(x_m, .., x_z)$ belonging to the union of the *ADI* and *PDI* sets. And it represents a **subset of the cartesian product** $D(x_m) \times ... \times D(x_z)$ that specifies allowed combinations of values for the variables $x_m...x_z$. The **result** of a *CAO* is an assignment v for that, each instance is a mapping that assigns to every variable $y_d \in PDO$ an element $v(y_d) \in D(y_d)$. This assignment v satisfies all the constraints belonging to C, such that $\langle\{y_{k_1}, ..., y_{k_g}\}, C_k\rangle \in C\ iff\langle v(y_{k_1}, ..., y_{k_g})\rangle \in C_k$ and optimize the global function.

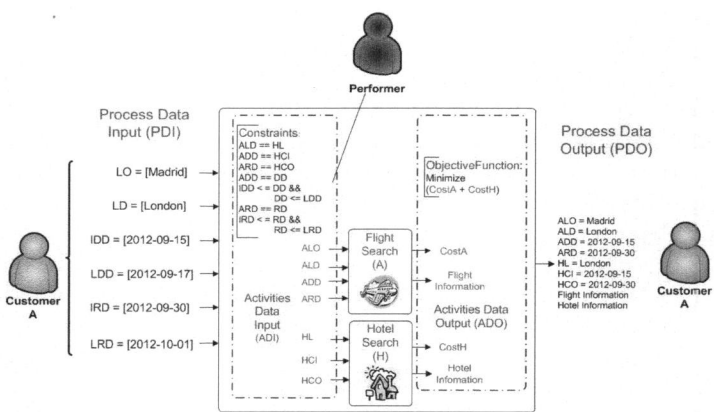

Fig. 1. The Travel Booking Example Structure

In the Travel Booking Example (see Figure 1), there are two activities: the Flight Search (A) and the Hotel Search (H). The *PDI* is the set of dates and locations given by the customer (location origin (LO), location destination (LD), initial depart date (IDD), last depart date (LDD), initial return date (IRD) and last return date (LRD)). The variables of the *ADI* are: airplane location origin (ALO), airplane location destination (ALD), airplane departure date (ADD), airplane return date (ARD), hotel location (HL), hotel check-in (HCI) and hotel check-out (HCO). There are many constraints that relate the *ADI* variables: the flight lands at the airport of the same city where is located the hotel, the departure and return date of the flight must match with the check-in and check-out date of the hotel, respectively. Moreover, there are constraints that relate

ADI with the *PDI*: for example, the value of the flight depart date has to fit with one of the depart dates given by the customer. On the other hand, there are two variables of the *ADO*: the cost of the flight and the cost of the hotel (CostA and CostH respectively). The objective function is to minimize the sum of both prices. Finally, the *PDO* is the set of the concrete values that optimize the objective function for the *ADI*.

4   BPMN Extension for Combining Activities That Involves Data Constraints

In business process modelling, one of the main goals is to facilitate the description of process model, shielding the user from unnecessary implementation details. However, BPMN 2.0 does not explicitly consider mechanisms to represent *CAO* as is defined in Section 3. For that reason, in order to capture these combination of activities within the business process, a notation supported by a set of graphical concepts is essential. This new notation enables a semantic representation and a graphical modelling. The extension presented in this section details this new notation and its marker associated.

BPMN 2.0 specification wants to stress the different stages in which the modeling process is composed: description, analysis and execution. The Description stage concerns the visible elements and attributes used in high-level modeling, in other words, the closest stage to the human level. The analysis stage contains all of the description stage and others, and it is closer to a software engine level. Both stages are focused on visible elements and a minimal subset of supporting attributes/elements. On the other hand, the execution stage focuses on what is needed to execute process models. To do this, following Subsections 4.1, 4.2, 4.3 and 4.4 define the metamodel, the new sub-process description, the operational semantics and the event handling, respectively. Finally, Subsection 4.5 presents a proposal to the executable stage. The typography, linguistic conventions and style of BPMN are also following to define this new extension.

4.1   Meta-model of the Combination of Activities Sub-Process

The Combination of Activities Sub-Process (CombA Sub-Process) meta-model proposed enables the analyst to describe and specify all aspects of *CAO* functionality and usage.

BPMN offers a way to represent the activities that have no REQUIRED sequence relationships through Ad-Hoc Sub-Process [7]. The Ad-Hoc Sub-Process semantics description is not enough for the combination of activities where the concrete values for the data input have to be found to optimize the goal, and there is no way to determine them at design time. Although there is no explicit process structure in Ad-Hoc Sub-Process, some sequence and data dependencies can be added to the details of the process. In addition, the performers[1]

---

[1] A Performer defines the resource that will perform or will be responsible for an activity. The performer can be specified in the form of a specific individual, a group, an organization role or position, or an organization [7].

determine when the activities will start, what the next activity will be, and so on. However, this is not valid when there are no sequence relationships, the data dependencies are on the data input of the activities and the performers cannot determine the sequence, as occurs in *CAO*. Furthermore, the list of BPMN elements that MUST NOT be used in an Ad-Hoc Sub-Process includes: Start and End Events, Conversations (graphically), Conversations Links (graphically) and Choreography Activities, which are useful to combine the activities to achieve an optimal goal.

In order to include these new requirements, an extension of Sub-Process definition [7] is necessary. A new contextual scope is defined to achieve an optimization agreement between different parts to search the data input values of the activities that optimize a goal. The new meta-model related to Sub-Process, adding the new type of Combination of Activities Sub-Process, called CombA Sub-Process, is presented in this paper.

4.2   CombA Sub-Process Definition

First of all, and according to the BPMN methodology, a descriptive stage is necessary to define the new sub-process. This description enables a high-level modeling through a visible element and a set of attributes.

CombA Sub-Process is a specialized type of Sub-Process which is a set of activities[2] that have no REQUIRED sequence relationships. This new sub-process has to combine the activities in order to search the concrete values for the interval data input handle that optimize a common objective. The set of activities can be defined in the process. However, the sequence and the number of performances for the activities cannot be determined by the performers of the activities, since the behavior of an activity depends on the data input belonging to other activities, and vice versa.

The main purpose of the combination of activities is to search the values of the data input that optimize a goal, not as much as the execution of the activities. However, the execution of the activities is a consequence since the objective function to optimize depends on the data output of the activities. Therefore, the activities have to be executed several times to agree which are the best values of the data input. If there were no intervals in the data input (since they have atomic values), the activities have to be executed only once to answer the customer request.

The formal definition presented in Section 3 is translated into BPMN elements in order to define the CombA Sub-Process. Therefore, the CombA Sub-Process is composed of:

– Activities (*A* in the formal definition): generally, these activities are tasks. A task is a unit of work, the job to be performed. However, each activity, in turn, can be another process. The unique requirement is that each activity has different and independent functionality.

---

[2] An Activity is a Process step that can be atomic (Tasks) or decomposable (Sub-Processes) [7].

- Data Object: it represents the information flowing through the process. The CombA Sub-Process handled mainly two types of data:
  - Data Input: there are two types of data input in the CombA Sub-Process: the first corresponds to the external input for the entire process. It can be read by an activity and is given by a performer (PDI in the formal definition); and the second type corresponds with the data input of each activity that participate in the optimization agreement (ADI in the formal definition).
  - Data Output: there are two types of data output: the first one corresponds to the variable available as result of the entire process and the answer of the customer request (PDO in the formal definition). And the second type corresponds to the output value of each activity (ADO in the formal definition).
- Constraints (C in the formal definition): a set of Formal Expressions[3] that relate both types of data input.
- Optimization Function: The Formal Expression used to define the function to be optimized. This function relates the data output of the activities.

As the visible element for the modelling, we propose to use a puzzle piece symbol like the marker for the CombA Sub-Process and it can be used in a collapsed and expanded way. The reason of choosing this symbol is that the activities in the CombA Sub-Process have to fit as the pieces of a puzzle, that can be a right visualization for the business level. The circular structure in the expanded CombA Sub-Process represents the lack of predefined order and sequence relationship, centered on the objective function, which is the CombA Sub-Process Core.

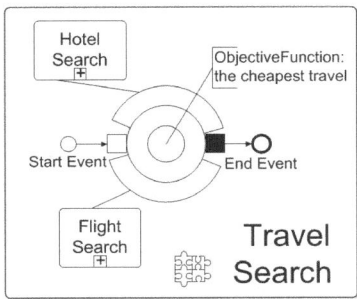

Fig. 2. Expanded Search Travel CombA Sub-Process

Figure 2 depicts the way to represent the combination of activities between Airline Searching and Hotel Searching activities with the new CombA Sub-Process marker.

---

[3] A Formal Expression is used to specify an executable Expression using a specified Expression language. A concrete constraint language is the one proposed in [6].

This CombA Sub-Process can be added to the diagram presented in [13] replacing the parallel search of flights and hotel rooms with this new Travel Search Sub-Process (see Figure 3).

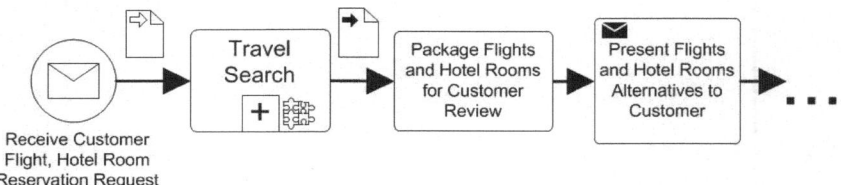

Fig. 3. Collapsed Travel Search CombA Sub-Process included in the example presented in [13]

The CombA Sub-Process element inherits the attributes and model associations of activities through its relationship to sub-process [7]. Table 1 presents the additional attributes and model associations of the CombA Sub-Process.

Table 1. CombA Sub-Process model associations

Attribute Name	Description - Usage
objectiveFunction: Formal Expression	Definition of the global optimization goal.
constraints: set of Formal Expressions	Definition of the constraints that relate the data input of the activities, limiting their possible behaviours.
numberSolutions: Integer	Attribute to determine if the CombA Sub-Process searches one of the best solutions or all the best solutions (values 0 or 1 respectively). The default value is 0.

Activities within this sub-process are generally disconnected from each other. During execution of the Process, all the activities are activated asynchronously and have to communicate their decisions to the others to achieve an agreement.

The formal definition given in Section 3 for the Trip Planner Example is valid for the CombA Sub-Process definition. Therefore, the attributes of CombA Sub-Process for the example are:

- **objectiveFunction:** To get the cheapest trip, the objective function is: $min(costA + costH)$.
- **numberSolutions:** All the best solutions regarding the objective function to optimize. According to the values given in Table 1, the value is 1.
- **constraints:** Two of the possible constraints of the problem are: the depart date has to be a value between the initial depart date and the last departure date given by the user (C1) and the destination airport is at the same location than the hotel (C2).

$$(C1) IDD \leq DD \ \&\& \ DD \leq LDD \quad (C2) ALD == HL$$

4.3   CombA Sub-Process Operational Semantics

As the definition of Sub-Processes presented in [7], a CombA Sub-Process is an activity that encapsulates a Sub-Process that is in turn modeled by Activities, Gateways, Events and Sequence Flow. Moreover, the activities involved in the combination could be composed of Conversations, Choreographies and other Sub-Processes. The CombA Sub-Process is instantiated when it is reached by a Sequence Flow token[4] through an unique Start Event. The CombA Sub-Process instance is completed when the activities achieve the optimal goal and there are no more tokens in the sub-process and none of its activities are still activated, in other words, when the End Event is reached.

As long as the activities do not find the optimal goal, and therefore, the concrete values for the data input, the sub-process will not end. The activities could find various best results since various combination of values can produce the same result. After the activities obtain an optimal value of the objective function, the value specified through the integer *numberSolutions* attribute is evaluated.

4.4   CombA Sub-Process Handling Events

One of the main problem of the combination of activities is the execution time. It can occur that the activities take a long time to find the combination of values that optimize the overall goal. In order to control this kind of problems, BPMN provides a set of Timer Events. If after a period of time no optimal goal is obtained, then the operation is canceled. Although, the CombA Sub-Process could return the best solutions found until that moment, it is not guaranteed that the optimal solution is among them. Sometimes, this timer requirement is provided by the customer who does not want to wait. However, timer requirement can also be used to provide quality products from the service provider point of view.

In general, BPMN provides several event handlers that help to manage and solve situations that happen during the course of a Process instance. Various of these event handlers can be applied to the CombA Sub-Process.

An Intermediate Event indicates where something happens (an Event), somewhere between the start and the end of a process. It will affect into the flow of the process, although this event will not start or (directly) terminate the process [7]. The Intermediate Timer Event, specifically the interrupting type, interrupts the activity, to which is attached, changing the normal flow into an exception flow.

Applied to the CombA Sub-Process (see Figure 4), it implies that there will be two outcomes: successful completion and failed completion.

Although in this paper only Timer Events are considered, a CombA Sub-Process can be combined with the set of Events given by BPMN 2.0 (e.g. Error,

---

[4] The concept of a token is used to facilitate the discussion of how Sequence Flows are used within a Process. A token will traverse the Sequence Flows and pass through the elements in the Process. A token is a theorical concept that is used as an aid to define de behavior of a Process that is being performed.

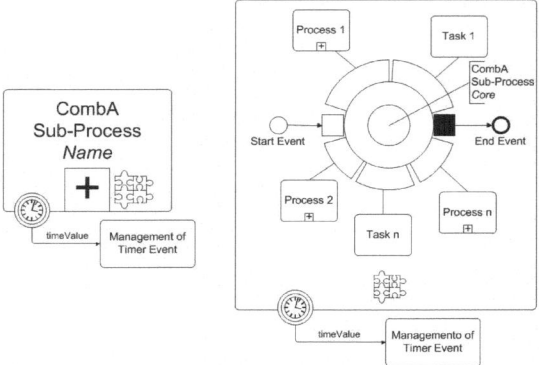

Fig. 4. A collapsed and expanded CombA Sub-Process with Timer Event

Message, Conditional, etc) [7] and completed handling these possible events as it is done by sub-processes. For example, an Error Event could warn when the problem is over-constrained. Since, although some over-constrained problems could be detected statically by the problem formulation, in the majority of cases it depends on the data returned by the activities. For example, if there are no flights for a concrete dates given by the user, then the problem is over-constrained.

4.5   CombA Sub-Process Execution Semantics

BPMN 2.0 pays special attention to the execution semantics since there is an important need of executable process models. In this subsection, a proposal is presented to implement and execute this CombA Sub-Process semantics.

A CombA Sub-Process contains a number of embedded inner activities and is intended to be executed with more flexible ordering, compared to the routing of Processes or Ad-Hoc Sub-Processes. The contained Activities in CombA Sub-Process are generally disconnected from each other and executed asynchronously. There are several ways to combine the activities and all of them depend on the performer (designer) criteria.

The combination of activities is formed by a set of activities whose objective is to search the concrete values of the data input that optimize an output. This problem is similar to Distributed Constraint Satisfaction Problems (DisCSP) [23], where the information is spatially and/or semantically distributed among various nodes where no one knows the whole information and the behaviour of each node. In [15], a methodology to model and execute combination of activities in business processes based on DisCSP is proposed. In order to solve the mentioned combination, it has developed an adaptation of the centralized backtracking algorithm existing for DisCSP. In a centralized solution, there is a coordinator that has the knowledge of the whole problem and ensures that the data input of the activities satisfy all the constraints and finds out the optimal value of the objective function. Thanks to this adaptation, the activities achieve the coordination to obtain an overall goal.

5   Related Work

One of the main ideas of BPMN is the imperative representation of business processes with their activities and the execution constraints between them. However, the execution order and the constraints that relate the activities cannot be always described with a traditional business process model.

There are many business process modelling techniques [1]: Event-Driven Process Chain (EPC) [17], UML [16] [9], Integration Definition for Function Modelling (IDEF) and Petri-Nets [3] [20]. Furthermore, Ruopeng Lu and Shazia Sadiq in [10] present a comparative of business process modelling approaches including graph-based modelling language and rule-based formalism. However, the most used graphical standard for modeling BP is Business Process Model and Notation (BPMN) [12] proposed by OMG, recently released the version 2.0 [7], which can be used for a wide range of problems [21]. This notation must also permit the incorporation of various perspectives giving place to various diagrams. The diagrams must show the rules, goals, objectives of the business and not only relationships, but also interactions [4]. A great part of the success of the modelling is the capacity to express the various needs of the business, as well as to have a notation in which these needs can be described. However, although it could not be an easy task, the business process, the environment features and the intended use of the model must be taken into account to make a successful choice of an approach and/or notation [2].

On the other hand, Yunzhou Wu et al. in [22] show how the constraints that necessitate coordination may be represented in the Business Process Execution Language for Web Services (BPEL). BPEL is an OASIS standard executable language for specifying actions within business processes with web services [11]. Processes in BPEL export and import information by using web service interfaces exclusively. But there is not a protocol defined in BPEL to combine the activities nor to select the concrete values of the data input according to the optimization function. The authors in [22] use a generalized adaptation and constraint enforcement models to transform the traditional BPEL process into an adaptive process. However, the authors solve the combined adaptation and constraint enforcement models in order by obtaining a policy that recommends adaptive actions while respecting the constraints. Therefore, there is no combination to search the concrete values for the interval data input in terms defined in our work. One way to represent the constraints is by means of the syntax of the Semantic Web Rule Language (SWRL) [19]. SWRL is used to specify the constraints and embed the constraints based on SWRL within BPEL.

Service-oriented systems have emerged as the paradigm to provide such automated support for business processes. Van der Aalst et al. in [18] and Papazoglou et al. in [14], present Web Services as the infrastructure to foster business processes by composing individual Web Services to represent complex processes.

There are several studies on the composition of services that can be extrapolated to the composition of activities in business processes since services are specific activities in the business process. To the best of our knowledge, none of these studies represents graphically or solves the type of coordination that this paper presents: a combination of independent activities to find the concrete values of their data input that optimize an overall objective.

Malinda Kapuruge et al. in [8] provide a systematic analysis of the requirements for process flexibility in the context of service compositions, and they analyze the existing approaches against these sets of requirements. Based on this analysis, some general observations are defined for certain critical issues for future investigation into business process flexibility support in service composition. However, a combination of the activities is possible against the lack of priority in the order of the activities presented in this paper.

Otherwise, Umeshwar Dayal et al in [5] define the coordination as a collaborative process, where the best service from among a set of existing and available services is chosen in order to fulfil a common need. Umeshwar Dayal et al. have analyzed and identified the requirements for business process flexibility in service composition and compared how existing process modelling and enactment approaches fulfil these requirements. However, in this work, each service that participates in the combination has an independent and distinct functionality. It is assumed that each of these services is the best at obtaining this functionality since the main objective in our work is that these services search the concrete values for their data input. Therefore, the way that these best services are chosen is irrelevant to this paper.

There are also various studies on planning algorithms for Web Services. The planning [24] has similar features with the same kind of combination as that presented in this paper. The input and output parameters of the distinct participating Web Services provide the basis of the planning. These relationships involve sequence relationships between the numerous Web Services. However, in this paper, there is no possibility of sequence relationships since only the data input are related.

6   Conclusions and Future Work

In this work, an extension of BPMN 2.0 for modelling the combination of activities that involve data constraints is proposed. The aim of this combination is to search the concrete values for the data handle by the activities in order to optimize an overall objective. This proposal arises from the need to combine various activities or sub-processes that belong to a business process and work concurrently achieving an optimization of an overall goal.

A CombA Sub-Process with its associated marker enables to incorporate combination requirements into a business process diagrams. These combination requirements will increase the scope of the expressive ability of business description. The CombA Sub-Process describes graphically the features of this combination of activities: there is a set of activities whose data input have a range of possible values, and have to optimize a common objective assigning concrete values to these data input. The main problem in this type of combination of activities is that the Activities have no required sequence relationships and their data input are related, by means of constraints, making possible several input and output data for each activity in an instance. There is no way to represent this with BPMN 2.0, hence the CombA Sub-Process element is proposed.

Future work will be oriented to enrich the combination of activities specifications and study in depth the analytic and common executable process modelling.

Acknowledgments. This work has been partially funded by the Junta de Andalucía by means of la Consejería de Innovación, Ciencia y Empresa (P08-TIC-04095) and by the Ministry of Science and Technology of Spain (TIN2009- 13714) and the European Regional Development Fund (ERDF/FEDER).

References

1. Aguilar-Saven, R.S.: Business process modelling: Review and framework. International Journal of Production Economics 90(2), 129–149 (2004)
2. Bidel, I.: Choosing approach to business process modeling - practical perspective (2007)
3. Bosilj-Vuksic, V., Hlupic, V.: Petri Nets and IDEF diagrams: Applicability and efficacy for business process modelling. An International Journal of Computing and Informatics 25(1), 123–133 (2001)
4. Castela, N., Tribolet, J.M., Silva, A., Guerra, A.: Business process modeling with uml. In: ICEIS (2), pp. 679–685 (2001)
5. Dayal, U., Hsu, M., Ladin, R.: Business process coordination: State of the art, trends, and open issues. In: Apers, P.M.G., Atzeni, P., Ceri, S., Paraboschi, S., Ramamohanarao, K., Snodgrass, R.T. (eds.) VLDB, pp. 3–13. Morgan Kaufmann (2001)
6. Gómez-López, M.T., Reina-Quintero, A., Martínez Gasca, R.: Model-Driven Engineering for Constraint Database Query Evaluation. In: First Workshop Model-Driven Engineering, Logic and Optimization: Friends or Foes? (MELO 2011) (June 2011)
7. Object Management Group. Business Process Model and Notation (BPMN) Version 2.0. OMG Standard (2011)
8. Kapuruge, M., Han, J., Colman, A.W.: Support for business process flexibility in service compositions: An evaluative survey. In: Australian Software Engineering Conference, pp. 97–106. IEEE Computer Society (2010)
9. List, B., Korherr, B.: A UML 2 Profile for Business Process Modelling. In: Akoka, J., Liddle, S.W., Song, I.-Y., Bertolotto, M., Comyn-Wattiau, I., Cherfi, S.S.-S., van den Heuvel, W.-J., Thalheim, B., Kolp, M., Bresciani, P., Trujillo, J., Kop, C., Mayr, H.C. (eds.) ER Workshops 2005. LNCS, vol. 3770, pp. 85–96. Springer, Heidelberg (2005)
10. Lu, R., Sadiq, S.: A Survey of Comparative Business Process Modeling Approaches. In: Abramowicz, W. (ed.) BIS 2007. LNCS, vol. 4439, pp. 82–94. Springer, Heidelberg (2007)
11. OASIS. Business Process Execution Language for Web Services. OASIS Standard (2005)

12. OMG. Business Process Model and Notation (BPMN) 1.2. OMG Standard (2009)
13. OMG. BPMN 2.0 by Example. Version 1.0 (non-normative). OMG Standard (2011)
14. Papazoglou, M.P., van den Heuvel, W.-J.: Service oriented architectures: approaches, technologies and research issues. VLDB J. 16(3), 389–415 (2007)
15. Parody, L., Gómez-López, M.T., Martínez Gasca, R., Borrego, D.: Using distributed csps to model business processes agreement in software multiprocess. In: 3rd International Conference on Agents and Artificial Intelligence (2011)
16. Sinogas, P., Vasconcelos, A., Caetano, A., Neves, J., Mendes, R., Tribolet, J.M.: Business processes extensions to uml profile for business modeling. In: ICEIS (2), pp. 673–678 (2001)
17. Tsai, A., Wang, J., Tepfenhart, W., Rosea, D.: Epc workflow model to wifa model conversion. In: Proceedings of IEEE International Conference on Systems, Man and Cybernetics, SMC 2006, vol. 4766, pp. 2758–2763 (2006)
18. van der Aalst, W.M.P., ter Hofstede, A.H.M., Weske, M.: Business Process Management: A Survey. In: van der Aalst, W.M.P., ter Hofstede, A.H.M., Weske, M. (eds.) BPM 2003. LNCS, vol. 2678, pp. 1–12. Springer, Heidelberg (2003)
19. W3C. SWRL: A Semantic Web Rule Language Combining OWL and RuleML. W3C (2004)
20. Weske, M.: Business Process Management: Concepts, Languages, Architectures. Springer (2007)
21. Wolter, C., Schaad, A.: Modeling of Task-Based Authorization Constraints in BPMN. In: Alonso, G., Dadam, P., Rosemann, M. (eds.) BPM 2007. LNCS, vol. 4714, pp. 64–79. Springer, Heidelberg (2007)
22. Wu, Y., Doshi, P.: Making bpel flexible - adapting in the context of coordination constraints using ws-bpel, pp. 423–430 (2008)
23. Yokoo, M., Durfee, E.H., Ishida, T., Kuwabara, K.: The distributed constraint satisfaction problem: Formalization and algorithms. IEEE Transactions on Knowledge and Data Engineering 10(5), 673–685 (1998)
24. Zhang, J.F., Kowalczyk, R.: Agent-based dis-graph planning algorithm for web service composition. In: CIMCA/IAWTIC, p. 258. IEEE Computer Society (2006)

Comparison of BPMN2 Diagrams

Pit Pietsch[1] and Sven Wenzel[2]

[1] University of Siegen, North Rhine-Westphalia, Germany,
`pietsch@informatik.uni-siegen.de`
[2] Technical University Dortmund, North Rhine-Westphalia, Germany,
`sven.wenzel@cs.tu-dortmund.de`

Abstract. Models are compared to identify which elements are unchanged and which were added, removed, or modified. This information is necessary for developers to understand which edit steps were applied between two revisions of a model, to discover differences in concurrently developed models and it is also a fundamental building block for advanced processing steps, e.g. model merging. Hence, model comparison is generally considered as a critical factor for the acceptance and success of model-driven development approaches. Surprisingly however, for many model types only inadequate tool support for comparing models is available. This paper presents the prototype of a similarity-based model comparison tool for BPMN2 diagrams. The algorithms and heuristics of the SiDiff model differencing framework have been configured to the specific characteristics of BPMN2 diagrams. An initial evaluation indicates that the presented prototype produces results of high quality.

Keywords: Model Comparison, BPMN2, Similarity of Business Process Models, Quality of Differences, Difference Computation.

1 Motivation

It is commonly acknowledged that model-based development must be supported by configuration management tools for models [1]. Especially tools for comparing and matching models, which we will collectively refer to as *differencing tools*, are a critical factor for the acceptance and success of model driven approaches in software development [2].

To understand in which regard two models differ is of major importance in the day-to-day routine of process modelers. For example, different versions of a model are compared to understand its evolution, models which stem from a common base are compared to understand the different development paths that were taken. Another example is the comparison of independently developed models that are to be merged and consolidated into a single model. Hence, model comparison is a fundamental building block for advanced processing steps. Surprisingly, there is still a significant lack of comprehensive differencing tools for various model types; model-type specific differencing tools are expensive in their development and thus not available for every domain, while generic differencing tools often produce results of low quality. The following examples illustrate such

J. Mendling and M. Weidlich (Eds.): BPMN 2012, LNBIP 125, pp. 83–97, 2012.
© Springer-Verlag Berlin Heidelberg 2012

low-quality correspondences produced by a generic model differencing tool. Figure 1 shows three Business Process Model and Notation Version 2.0 (BPMN2) diagrams[1] which model a simple shipment process.

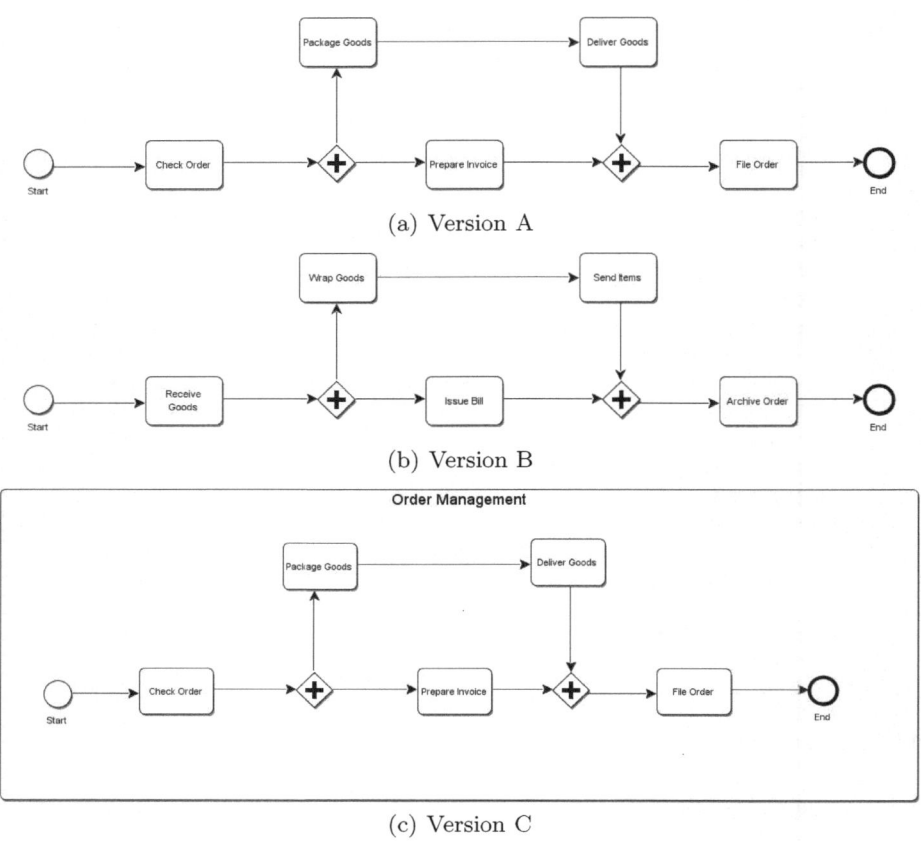

(a) Version A

(b) Version B

(c) Version C

Fig. 1. Variants of a simple shipment process

Between Version A and Version B there are no structural changes, i.e. no edit steps that add or delete model elements were applied; instead all tasks in Version B have been renamed. Version C again models the same process as Version A, only that it has been moved into the new subprocess *Order Management*. The models are compared with EMF Compare [7], a well-known generic model differencing tool integrated in the Eclipse IDE. It is assumed for the following examples that EMF Compare is configured to compare models only on the basis

---

[1] All models presented in this paper comply to the implementation of the BPMN2 standard [4] provided by the Eclipse BPMN2 project [5]. The diagrams were created with the BPMN2 Visual Editor for Eclipse [6].

of similarities, i.e. persistent identifiers are either not available in the models, or the models were developed independently or with different tools.

Figure 2 shows the computed difference between Version A and Version B. Only one renaming, i.e. the renaming of task *Check Order* to *Receive Order*, is recognized correctly. The other four renamings were not recognized and instead deletions and creations of the tasks are reported. Furthermore all sequence flows were matched, even if their source and target elements are reported as not corresponding. Arguably, the computed difference is of low quality and does not represent the actual evolution.

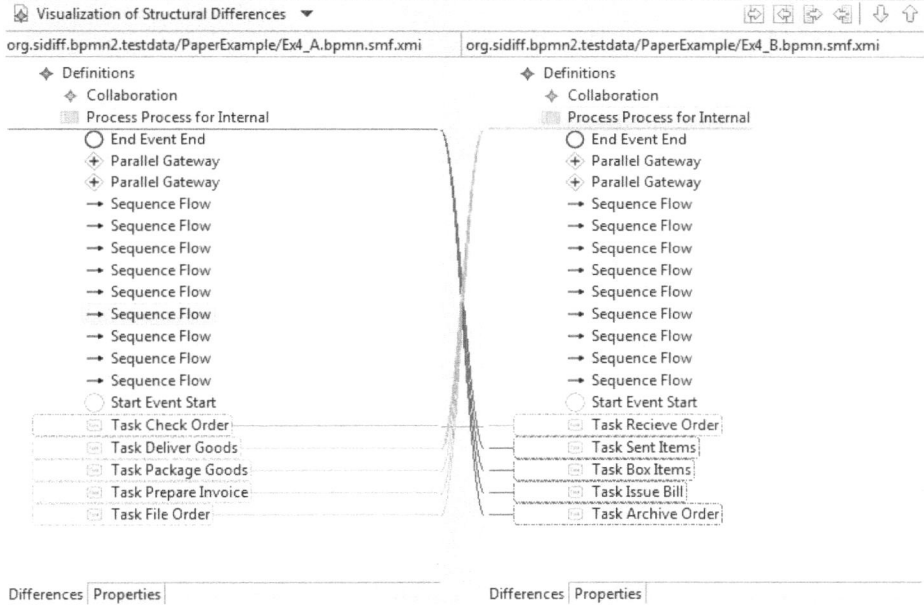

Fig. 2. Difference between version A and B reported by EMF Compare

Figure 3 shows the correspondences computed during the comparison of Version A and Version C. The move of the original process into the newly added subprocess was not recognized, and thus all elements are reported as deleted from Version A. Accordingly, the tool reports for Version C, that the subprocess *Order Management* and all the contained tasks have been created (instead of being moved).

Arguably, in both cases the delivered difference is of low quality. As will be discussed in Section 3, the cause for these poor results lies in the heuristics used by the matching algorithms of EMF Compare.

This paper presents the prototype of a similarity-based model differencing tool for BPMN2 diagrams. The tool has been realized by configuring the SiDiff model differencing framework [8,9]. SiDiff consists of a set of highly configurable

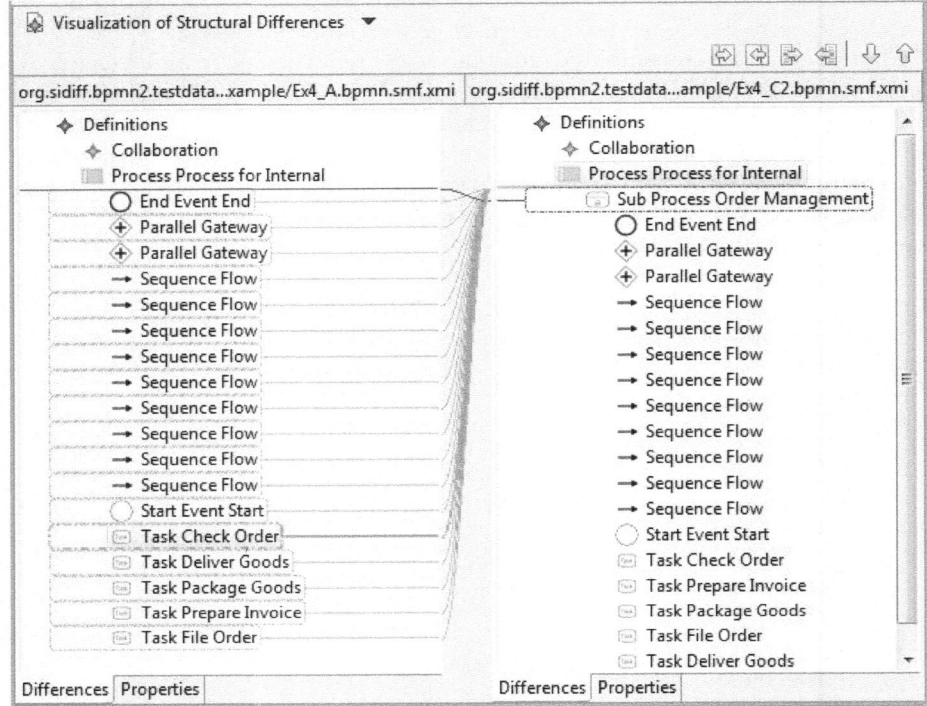

Fig. 3. Difference between version A and C reported by EMF Compare

matching algorithms; the heuristics used in the model comparison process are
custom-tailored in order to fit to the characteristics of BPMN2 diagrams.

The rest of the paper is structured as follows: Section 2 gives a summary
of state-of-the-art model comparison approaches. Section 3 introduces the basic
principles behind the SiDiff framework and excerpts of the configuration for
BPMN2 diagrams are presented. In Section 4 the prototype is evaluated and the
quality of the computed matchings is discussed. Related work and future work
are presented in Section 5 and Section 6, respectively. Finally the paper closes
with a summary of its contents in Section 7.

2   Background Model Comparison

Generally, models are compared to identify which of the contained elements are
unchanged and which were added, removed, or modified. Therefore a differencing
tool must primarily deliver two things:

- a *matching*, i.e. all pairs of corresponding model elements.
- a *difference*, i.e. a sequence of edit steps which converts the first model into
  the second.

The difference can be easily derived from the matching. Model elements which are not involved in a correspondence are reported to be either deleted or created; elements that do correspond, but have different properties in the two models, are reported as updated; finally, elements are reported as unchanged if they correspond and are identical in their properties. The further processing of the difference depends on the particular use case, e.g. difference visualization, patching, merging, etc. The most crucial factor for the quality of the difference is the quality of the previously computed matching.

In the last decade a large number of differencing tools were proposed in the literature[2]. Some differencing tools are model type specific. The heuristics used by their comparison algorithms are custom-tailored to fit the characteristics of one specific model type. The quality of the produced differences is generally high. Other approaches are generic in the sense that they are applicable to arbitrary model types. The quality of differences computed by generic approaches depends to a large extent on the configurability of their algorithms and the quality of the configuration data that is provided for a given model type. Generic differencing tools can be divided further into signature-based and similarity-based approaches.

Signature-based approaches match only elements that have an identical signature. In practice, the signature of an element is usually a value that is based on one or more of its properties. Examples of signature are fully qualified names, hash values of elements or persistent identifiers. Correspondence computation based on signatures is a simple and efficient two-step process:

1. The signature values for all elements of each model are computed and sorted.
2. Values contained in both signature sets are identified.

A correspondence between two model elements is established if they have identical signature values and the value is unique for each of the two sets. No correspondence is established in the case of an ambiguity, i.e. a signature value is contained more than once in one of the sets. The execution time for signature-based model differencing is in the order of $O(n*log(n))$, the most dominant part being the sorting of the signature sets.

Similarity-based differencing approaches try to match the mutually most similar model elements, while these elements are not necessarily identical. The configuration data for these algorithms defines how a similarity value for two model elements is computed. Their overall similarity is usually based on one or more properties that are considered as relevant for their similarity.

The matching process of similarity-based approaches consists of three steps and is more complex compared to the two-step process of signature-based approaches.

1. The similarity for each pair of model elements is computed.
2. If the similarity between two elements exceeds a threshold, then both elements are considered as candidates for a correspondence. For each model

---

[2] Surveys can be found in [10,11,12].

element, all candidates in the other model are collected in a preference list, sorted by their similarity.
3. A matching algorithm computes all correspondences on the basis of the preference lists.

The execution time of the first step is in the order of $O(n^2)$, which makes similarity-based model differencing runtime-expensive when large models are processed.

3   A Model Differencing Tool for BPMN2 Diagrams

As discussed previously, the quality of the derived difference depends mainly on the quality of the initially computed matching. The quality of the computed matching again depends on the heuristics used by the model comparison algorithms. Especially how these heuristics relate to the characteristics of a given model type is an important factor here. It is generally acknowledged that there is not one single optimal heuristics which can be used in all application scenarios and all model types [11,35]. Inappropriate heuristics on the other hand often lead to poor decisions when model elements are matched.

For example, the reason for the low quality of the difference between Version A and Version B discussed in Section 1 is that the EMF Compare matching algorithm puts a strong emphasis on the names of elements. This assumption works well for class diagrams and similar model types, where almost all model elements have names and can often even be uniquely identified based on them. Conversely, this heuristic fails for model types where elements often have no or only generic names. BPMN2 diagrams are even more complicated, because they often contain unnamed elements that also lack additional local properties which can be used in the similarity computation, e.g. *Sequence Flows*. Hence, model differencing tools for process models have to take the environment of such elements into account to perform a reliable matching.

The poor quality of the computed difference in the second example, where the moved tasks are not recognized, can be explained by another inappropriate heuristic that makes assumptions about the structure of a model; EMF Compare uses a *top-down* matching strategy. The matching starts at the root elements of the models. Subsequently only the direct children of the root elements are compared and matched and so on. This way the matched elements always form a tree. A top-down matching heuristic is efficient, because the set of compared elements is strictly limited. However, based on a top-down matching moved model elements can not be detected at all. Again, this a reasonable optimization for structural model types like class diagrams where moves seldom occur, but it leads to poor matching decisions if BPMN2 diagrams or other models types with recursive structures are compared.

Hence, in order to compute high quality correspondences for BPMN2 diagrams, the environment of a model element has to be taken into account by the matching algorithm. Furthermore, the used heuristics must not assume hierarchies as the primary concept for structuring. Since an element, e.g. a task, might be moved from one subprocess to another, the heuristics rather have to be able to recognize moved elements. These requirements can only be fulfilled by using a *bottom-up global* matching strategy. This heuristic searches globally for potential correspondences [15,9]. All pairwise similarities between model elements of the same type are computed before any element pair is matched. This allows to propagate the computed similarities [16]. Hence, the similarity between the neighbourhoods of two compared model elements can here be added to their basic similarity. Due to cyclic dependencies, similarity propagation is an iterative, compute-intensive process which is not guaranteed to converge. Different optimization strategies which generally achieve acceptable runtime performance exist, though [9,17].

The SiDiff model differencing framework is especially designed to be adaptable to arbitrary model types. The comparison algorithms of the framework can be tightly coupled to the specific characteristics of a model type and their heuristics can be individually configured for each model element type. The remainder of this section gives a short summary[3] of the important SiDiff features which are used to compare BPMN2 diagrams. Generally, SiDiff is adapted to a model type based on two configurations: the *similarity configuration* and the *matching configuration*.

Table 1. Simplified similarity configuration for tasks

Metaclass	Task
Similarity Threshold	0.3
Criterion/condition	**Weight**
Similar name	0.3
Similar outgoing sequence flows	0.15
Similar incoming sequence flows	0.15
Similar remote neighbours incoming	0.2
Similar remote neighbours outgoing	0.2

Similarity Configuration. The similarity configuration defines for each element type a set of similarity-relevant properties, a similarity function for each of these properties, and a weight. Relevant properties can either be local, e.g. the name of an element, or remote, e.g. similarities of neighbour elements. The similarity function returns a value between 0.0 and 1.0 for two given property values. A similarity of 0.0 indicates no similarity, while a value of 1.0 indicates that the properties have identical values. Obviously, some properties of elements are more important for their similarity than others. Therefore an additional weight is assigned to the similarity of each property. Thus, the total similarity between two elements is defined as the weighted mean of the similarities of all relevant properties. Furthermore, a similarity threshold defines the minimum similarity

---

[3] For an in depth discussion of the framework and its configuration possibilities see [8,9].

necessary for elements to be deemed as similar. If the similarity is below this threshold, a similarity of 0.0 is assumed.

An simplified example of a configuration for the element type *Task* is presented in Table 1. As can be seen, the similarity of two task is determined based on their names, the similarities of incoming and outgoing sequence flows and the similarities of the elements that are connected through these sequence flows.

Matching Configuration. When and how elements are matched is defined in the matching configuration. An excerpt of the configuration for the element type *Task* can be seen in Table 2. The *matching threshold* defines the minimum similarity for elements to be considered as eligible for a correspondence. The option *movable* defines whether or not elements can be matched if they have different parent elements, e.g. if they were moved in a different subprocess of the model. Prohibiting moves effectively reduces the search space and speeds up the matching computation, although it can diminish the quality of the computed matching if used inappropriately. The similarity for tasks depends partially on the similarity of neighbor elements. These similarities may change in each iteration of the algorithm. Hence, task are considered to be *part of the similarity flooding*, i.e. their similarity must also be computed in each iteration. Accordingly, similarity values for elements that are not *part of the similarity flooding* are calculated only once at the beginning of the matching process.

Table 2. Excerpt from the matching configuration for Tasks

Meta Class	Task
Matching Threshold	0.5
Option	**Value**
Moveable	true
Part of similarity flooding	true
...	

Obviously, the matching and the similarity configurations are strongly interwoven. Therefore the matching thresholds, as well as the similarity thresholds and the weights assigned to the similarity functions must be selected with great care.

4   Evaluation

The most important assessment criteria for the quality of the presented model differencing prototype for BPMN2 diagrams is the quality of the computed matching. The evaluation is based on three different data sets which contain a total of 19 different BPMN2 diagrams. All processes models were created by hand. Hence, all edit steps applied during the modification process as well as the correct correspondences are known. Accurate precision and recall values are easily deducible based on the correct and calculated correspondences.

Table 3. Data Set II

Testcase	Precision	Recall
A_01 ↔ A_02	0.80	0.89
A_02 ↔ A_03	0.90	1.00
A_03 ↔ A_04	0.92	0.96
A_04 ↔ A_05	0.94	1.00
A_05 ↔ A_06	0.95	1.00
A_06 ↔ A_07	0.96	0.96
A_07 ↔ A_08	0.96	1.00
A_08 ↔ A_09	1.00	1.00

Data Set I. The first data set contains the BPMN2 diagrams introduced in Section 1. For both test cases the matching computed by SiDiff proved to be completely correct. Thus, the precision and recall values for this data set are 1.0, respectively.

Data Set II. The second data set models the shipment process of a hardware retailer. This test case is based on the example process discussed in Section 5.1 of [18]. The history of this model spans a total of 9 revisions. The first revision contains 11 different model elements, while the last revision consists of 54 elements. On average, a model in this data set has around 38 elements. The average precision and recall values are 0.930 and 0.975, respectively. The results for each pair of models are presented in Table 3.

Data Set III. The last data set is again based on an example from [18]; this time a collaboration process mimicking a pizza delivery service is modeled. Just like Data Set II, this data set was also iteratively developed and the history spans seven different revisions. The smallest revision consists of 33 elements, while model representing the largest model contains 71 elements. On average a model of this data set contains about 46 elements. The quality of the computed matching is with an average precision of 0.990 and an average recall of 0.963 comparable to the results achieved in the evaluation of Data Set II. The details of the evaluation are presented in Table 4.

Table 4. Data Set III

Testcase	Precision	Recall
A_01 ↔ A_02	1.00	0.97
A_02 ↔ A_03	1.00	0.97
A_03 ↔ A_04	0.95	0.95
A_04 ↔ A_05	1.00	0.98
A_05 ↔ A_06	1.00	0.98
A_06 ↔ A_07	0.98	0.94

5   Related Work

In the literature, first attempts to compare process models have been proposed. They are discussed in the following subsection. Due to the use of a configurable differencing approach, other generic differencing tools are discussed afterwards. Finally, recording based approaches are briefly discussed, as they provide an alternative to model differencing.

5.1   Comparison of Processes Models

Differences of Process Models. Dijkman et al. [19,20] discuss the problem of differencing and merging similar, but independently developed process models. It is suggested in both papers that equivalences of functions must be established manually first. Such manual matching approaches might be feasible for application scenarios where the matching has to be performed only once or when the differences between models is expected to be large, e.g. when process models repositories are merged. However, they are impractical in daily development routines where model differencing has to be performed on a regular basis.

Another approach that computes differences between process models has been presented by Li et al. [32]. Here, the distance between process models is measured based on the minimal edit script which transforms one model into the other. The algorithm and its heuristics are not configurable and is it also assumed that the activities in the process models are uniquely labeled.

Merging of Process Models. Küster et al. discuss in [21] how model differences can be detected and resolved when the application of edit steps is not logged. Correspondences are assumed as given based on unique element identifiers. Gerth et al. [22] discuss a language independent approach to detect conflicting change operations in process models. Here it is assumed that the correspondences were already computed with the help of existing model comparison algorithms.

Merging of models and resolving the conflicts that arise during the process are advanced processing steps that are based on model comparison. Obviously, high quality matchings and differences are essential to perform high quality model merging. Hence, the presented prototype of a differencing tool is complementary to this line of research.

Similarity Search of Process Models. Several papers, e.g. [25], [23], [24], discuss the problem of similarity search of process models. Here, the task is to find the most similar process model(s) out of a repository for a given query model or fragment thereof. Several graph matching algorithms which compute the similarity of whole process models based on the graph structured, the execution semantic and the labels of elements are proposed as solutions. The general problem of these algorithms is their complex runtime behaviour, some are even NP-complete. This makes their application impractical for large models and large repositories. An approach which displays a better runtime behaviour was proposed recently by Weidlich et al. in [36]. The general problem of similarity search is a research topic of its own merit and is not in the scope of this paper.

5.2   Generic Model Comparison Tools

Lin et al. [14] match model elements on the basis of signatures which are deterministically computed from the properties of model elements. A major problem of signature-based approaches is that signatures often encode the names of elements, which prohibits the matching of renamed elements. The approach furthermore uses a top-down matching strategy. If two elements cannot be matched, all elements belonging to the subtrees below these unmatched elements cannot match, either. Structural properties are hardly considered.

Xing and Stroulia use heuristics that are applicable to structural models [13], however, they are significantly tailored to the paradigm of object orientation. Models with flow characteristics such as BPMN2 are not supported, as one can see in the evaluation of the EMF Compare tool [7] which is built upon the ideas of Xing and Stroulia.

Melnik et al. presented a graph matching algorithm, which is based on the idea that the similarity between a pair of vertices is determined by the similarity between its neighbors [16]. This approach was a basis for the heuristics we developed here, however, Melnik et al. work on directed labeled graphs and do not consider domain-specific knowledge.

5.3   Recording Based Approaches

Another way to describe the evolution of a model is based on the sequence of the applied edit steps. Basically, such recording approaches, e.g. [26,27,28] maintain a log file which records all edit steps that are applied to a model. Hence, it is implicitly assumed that each editor and all tools which are used to modify the model can also access and modify the information maintained in the log file. Model differencing based on logging is efficient, but can only be applied under specific conditions. The most noteworthy is that the models must be in a direct predecessor-successor relationship. It is not possible to compare models which stem from different sources with recording-based approaches. Furthermore, these approaches prohibit concurrent modifications of a model. These constraints are often not met in today's heterogeneous open tool environments and distributed development processes.

6   Future Work

This paper presented the prototype of a model differencing tool for BPMN2 diagrams. While first results are promising, the prototype is a work in progress and needs to be further optimized before it can be used in real development environments. We currently plan to advance the prototype in three different aspects.

Advanced Heuristics. The currently used heuristics for matching BPMN2 diagrams are based only on the computation of similarities and their propagation. Additional efficient heuristics, e.g. signature-based matching algorithms, are not implemented in the prototype. Thus, we plan to adapt and incorporate these efficient heuristics to speed up the comparison process.

Advanced Evaluation and Configuration Optimization. Currently we do not have access to real BPMN2 diagrams for evaluation purposes. Hence, the evaluation in Section 4 is based on sets of rather small, handish created test models. While this first evaluation produced very good results, an evaluation based on large, realistic models is necessary to predict the quality of calculated differences more accurately. It is also likely that the configurations discussed in Section 3 must be re-evaluated and adjusted based on the results of such an extensive evaluation. Furthermore, large test models are also necessary to assess the average runtime behaviour of the prototype. Another possibility to obtain realistic test models is to synthetically create them with a model generator, e.g. [33,34].

Semantic Lifting. The differences reported by the prototype contain only simple, low-level edit steps on the graph representation of the BPMN2 diagrams. These simple edit steps do not always make sense from the perspective of a user, because users are accustomed to edit models based on more complex operations, e.g. refactorings. Thus, the currently computed difference should be semantically lifted to the level of user edit operations, as proposed in [29]. In the future we plan to identify user-level edit operations on BPMN2 diagrams and to semantically lift the current low-level difference so that the computed results are easier to comprehend.

7 Summary and Conclusion

The task of comparing two models is of major importance in the day-to-day routine of process modelers; which elements in two versions of a model correspond, were changed, added or deleted is essential knowledge to understand its evolution. Model comparison is also fundamental for many advanced model versioning functionalities, e.g. model merging and patching. Thus, model differencing tool should support process modelers and compute matchings and differences of high quality. However, common generic differencing tools generally perform poorly and produce low-quality results when process models are compared.

This behaviour was shown by example in Section 1 when three BPMN2 diagrams modeling a simple process were compared. The computed correspondences and differences were generally of low-quality and are in all likelihood perceived as wrong by process modelers. The root causes for this behaviour were identified and discussed: The algorithms of state-of-the-art generic differencing tools are well optimized for structural model types like class diagrams, but their heuristics neither are, nor can be adapted to sufficiently support other model types, notably model types with flow characteristics like BPMN2 diagrams.

To answer the presented deficiencies of generic state-of-the-art differencing tools, a prototype of a BPMN2 differencing tool was outlined in Section 3 . The prototype is based on the SiDiff model differencing framework. The most distinctive feature of SiDiff is that the framework is highly adaptable, i.e. the heuristics used by the model comparison algorithms can be custom-fit to the characteristics of arbitrary model types. It was discussed in detail which heuristics of state-of-the-art tools are responsible for the low-quality results and in contrast how SiDiff can be configured to properly address the specific characteristics of BPMN2 diagrams.

The presented prototype implementation was evaluated in regard to the quality of the computed correspondences in Section 4. The results achieved in this initial evaluation are promising. The average precision values varied between 0.93 and 1.0 and the average recall values varied between 0.963 and 1.0, depending on the data set.

Still, further test and optimizations have to be performed before the prototype can be applied within real development environments or incorporated into existing business process management tools. As discussed in Section 6 this includes evaluating and optimizing the currently implemented heuristics based on real models. Furthermore, additional heuristics should be implemented to optimize and speed up the model comparison process. This is necessary to assure that the runtime of the presented prototype is still acceptable when large models are compared. Both mentioned issues are in the focus of current research.

Acknowledgements. This work of the first author was supported by the QuDiMo project [30] of the German Research Foundation (Deutsche Forschungsgemeinschaft, DFG) under grant KE499/5-1 and the work of the second author was supported by the SecureClouds project of the BMBF under grant 01IS11008D.

References

1. Bendix, L., Emanuelsson, P.: Collaborative work with software models - industrial experience and requirements. In: International Conference on Model-Based Systems Engineering, MBSE 2009, pp. 60–68 (March 2009)
2. Bendix, L., Emanuelsson, P.: Diff and merge support for model-based development. In: Proceedings of the 2008 International Workshop on Comparison and Versioning of Software Models, CVSM 2008, pp. 31–34. ACM, New York (2008)
3. Gerth, C., Küster, J.M., Luckey, M., Engels, G.: Precise Detection of Conflicting Change Operations Using Process Model Terms. In: Petriu, D.C., Rouquette, N., Haugen, Ø. (eds.) MODELS 2010, Part II. LNCS, vol. 6395, pp. 93–107. Springer, Heidelberg (2010)
4. OMG, Business Process Model and Notation (BPMN) Version 2.0 (2011), http://www.omg.org/spec/BPMN/2.0/PDF (accessed May 15, 2012)
5. Eclipse BPMN2 Project (2012), http://www.eclipse.org/modeling/mdt/?project=bpmn2 (accessed May 15, 2012)
6. Eclipse BPMN2 Visual Editor (2012), https://git.eclipse.org/c/bpmn2-modeler/org.eclipse.bpmn2-modeler.git/ (accessed May 15, 2012)
7. EMF Compare Project (2012), http://www.eclipse.org/emf/compare (accessed May 15, 2012)

8. Kelter, U., Wehren, J., Niere, J.: A generic difference algorithm for uml models. In: Software Engineering 2005. Fachtagung des GI-Fachbereichs Softwaretechnik (2005)
9. Treude, C., Berlik, S., Wenzel, S., Kelter, U.: Difference computation of large models. In: ESEC-FSE 2007: Proceedings of the the 6th Joint Meeting of the European Software Engineering Conference and the ACM SIGSOFT Symposium on the Foundations of Software Engineering, pp. 295–304. ACM, New York (2007)
10. Förtsch, S., Westfechtel, B.: Differencing and merging of software diagrams - state of the art and challenges. In: ICSOFT (SE), pp. 90–99 (2007)
11. Kolovos, D.S., Ruscio, D.D., Pierantonio, A., Paige, R.F.: Different models for model matching: An analysis of approaches to support model differencing. In: CVSM 2009 (2009)
12. Selonen, P.: A review of UML model comparison techniques. In: Nordic Workshop on Model Driven Engineering (2007)
13. Xing, Z., Stroulia, E.: UMLdiff: An algorithm for object-oriented design differencing. In: ASE 2005: Proceedings of the 20th IEEE/ACM International Conference on Automated Software Engineering, pp. 54–65. ACM, New York (2005)
14. Lin, Y., Gray, J., Jouault, F.: DSMdiff: A differentiation tool for domain-specific models. European Journal of Information Systems 16(4), 349–361(13) (2007)
15. The SiDiff Project - Similarity-based differencing of models (2012), http://sidiff.org (accessed May 15, 2012)
16. Melnik, S., Garcia-Molina, H., Rahm, E.: Similarity flooding: A versatile graph matching algorithm and its application to schema matching. In: 18th Intl. Conf on Data Engineering, ICDE (2002)
17. Pietsch, P.: Optimierung des SiDiff-Algorithmus unter Ausnutzung modellspezifischer Eigenschaften und Strukturen. Master's thesis, University of Siegen (2008)
18. OMG, BPMN 2.0 by Example, http://www.omg.org/spec/BPMN/20100601/10-06-02.pdf (accessed May 15, 2012)
19. Dijkman, R.M.: A classification of differences between similar business processes. In: EDOC, pp. 37–50 (2007)
20. Dijkman, R.: Diagnosing Differences between Business Process Models. In: Dumas, M., Reichert, M., Shan, M.-C. (eds.) BPM 2008. LNCS, vol. 5240, pp. 261–277. Springer, Heidelberg (2008)
21. Küster, J.M., Gerth, C., Förster, A., Engels, G.: Detecting and Resolving Process Model Differences in the Absence of a Change Log. In: Dumas, M., Reichert, M., Shan, M.-C. (eds.) BPM 2008. LNCS, vol. 5240, pp. 244–260. Springer, Heidelberg (2008)
22. Gerth, C., Küster, J.M., Engels, G.: Language-Independent Change Management of Process Models. In: Schürr, A., Selic, B. (eds.) MoDELS 2009. LNCS, vol. 5795, pp. 152–166. Springer, Heidelberg (2009)
23. Dumas, M.: Similarity search of business process models (2009)
24. Dijkman, R., Dumas, M., García-Bañuelos, L.: Graph Matching Algorithms for Business Process Model Similarity Search. In: Dayal, U., Eder, J., Koehler, J., Reijers, H.A. (eds.) BPM 2009. LNCS, vol. 5701, pp. 48–63. Springer, Heidelberg (2009)
25. van Dongen, B.F., Dijkman, R., Mendling, J.: Measuring Similarity between Business Process Models. In: Bellahsène, Z., Léonard, M. (eds.) CAiSE 2008. LNCS, vol. 5074, pp. 450–464. Springer, Heidelberg (2008)

26. Herrmannsdoerfer, M., Koegel, M.: Towards a generic operation recorder for model evolution. In: Proceedings of the 1st International Workshop on Model Comparison in Practice, IWMCP 2010, pp. 76–81. ACM, New York (2010)
27. Lippe, E., van Oosterom, N.: Operation-based merging. In: Proceedings of the Fifth ACM SIGSOFT Symposium on Software Development Environments, SDE 5, pp. 78–87. ACM, New York (1992)
28. Schneider, C., Zündorf, A., Niere, J.: Coobra – A small step for development tools to collaborative environments. In: Workshop on Directions in Software Engineering Environments; Workshop at ICSE 2004 (2004)
29. Kehrer, T., Kelter, U., Taentzer, G.: A rule-based approach to the semantic lifting of model differences in the context of model versioning. In: ASE, pp. 163–172 (2011)
30. Pietsch, P., Yazdi, H.S.: The QuDiMo Project (2011), http://pi.informatik.uni-siegen.de/qudimo/ (accessed June 26, 2012)
31. Polyvyanyy, A., Vanhatalo, J., Voelzer, H.: Simplified computation and generalization of the refined process structure tree. In: Proceedings of the 7th International Workshop on Web Services and Formal Methods, WS-FM (2010)
32. Li, C., Reichert, M., Wombacher, A.: On Measuring Process Model Similarity Based on High-Level Change Operations. In: Li, Q., Spaccapietra, S., Yu, E., Olivé, A. (eds.) ER 2008. LNCS, vol. 5231, pp. 248–264. Springer, Heidelberg (2008)
33. Pietsch, P., Shariat Yazdi, H., Kelter, U.: Generating realistic test models for model processing tools. In: ASE 2011, pp. 660–623 (2011)
34. Pietsch, P., Shariat Yazdi, H., Kelter, U.: Controlled Generation of Models with Defined Properties. In: SE 2012 (2012)
35. Kehrer, T., Kelter, U., Pietsch, P., Schmidt, M.: Adaptability of Model Comparison Tools. In: ASE (2012)
36. Weidlich, M., Mendling, J., Weske, M.: Efficient consistency measurement based on behavioral profiles of process models. Transactions on Software Engineering 37(3), 410–429 (2011)

A Tool for Animating BPMN Token Flow

Thomas Allweyer and Stefan Schweitzer

Fachhochschule Kaiserslautern, Fachbereich Informatik und Mikrosystemtechnik,
Amerikastr. 1, 66482 Zweibrücken, Germany
thomas.allweyer@fh-kl.de, stsc0048@stud.fh-kl.de

Abstract. The concept of tokens flowing through a process model is very useful for explaining and understanding the meaning and the execution semantics of a BPMN model. This paper presents a software tool for animating the token flow of arbitrary process models. It can handle different scenarios of gateway combinations, loops, expanded and attached sub-processes, untyped start and end events, as well as terminating end events. It is possible to show several process instances within the same model. They are represented as differently colored tokens.

Keywords: Animation, BPMN, E-Learning, Execution Semantics, Sequence Flow, Token Flow.

1 Introduction

The semantics of BPMN sequence flow can be explained by the concept of tokens flowing through the model. The BPMN specification introduces tokens as a theoretical concept "as an aid to define the behavior of a process" [1]. The instantiation of a process is represented by the start event producing a token. This token then travels through the sequence flow until it is consumed by an end event. When it reaches an exclusive split, the conditions of each alternative flow are evaluated, and the token is routed to exactly one of these alternative flows. When a token arrives at a splitting parallel gateway, it is duplicated, and the outgoing sequence flows receive one token each. A joining parallel gateway will emit one token after all parallel tokens have arrived; i. e. after each entry has received a token. Possible misinterpretations of the meaning of parallel flows, such as the need for simultaneous processing or simultaneous completion of parallel activities, can be corrected very easily by explaining the behavior of the tokens.

Other BPMN elements, such as inclusive gateways or sub-processes, can also be defined very precisely with the token flow concept. This concept has also proved to be very useful for teaching and learning BPMN. Several BPMN books use token flows for explaining the meaning of process models, e.g. [2], [3], [4], [5].

The token flow concept represents the execution semantics of BPMN models, i. e. the token flow describes how a process engine would execute the model. Although many models are not executable, every BPMN model should be not only syntactically correct, but also semantically. The sequence flow should correctly

J. Mendling and M. Weidlich (Eds.): BPMN 2012, LNBIP 125, pp. 98–106, 2012.

reflect the modeler's intentions of how the process is actually performed. Ambiguities, different interpretations etc. can be resolved, if the token flow is analyzed – and errors can be detected, such as wrong gateway types.

As the token flow is a theoretical concept, its analysis is usually only done mentally. The analyst just imagines how tokens are created, routed, duplicated, merged and consumed. Such analyses are neither carried out systematically, nor regularly. For executable processes, modeling errors may be detected during testing the implemented process. However, testing does not detect all errors, and it would be cheaper if modeling errors could be detected earlier.

For non-executable processes, such modeling errors may not be detected at all. There are modeling tools which help detecting syntactical modeling errors. It is also possible to automate the analysis of the execution behavior in order to find problems, such as dead locks [6, p. 267 ff]. Modeling errors that are much harder to detect are deviations of the model's actual behavior from the intended behavior. Systematic token flow analyses could help the modeler recognizing whether the model behaves as expected.

Such a systematic token flow analysis can be supported by visually animating the token flow. Especially for beginners it is difficult to mentally "play through" a model. Therefore, many trainers use animated presentation slides in BPMN courses for visualizing the token flow of simple models, and there are also some web sites containing process model animations, e.g. [7], [8].

In this paper we present a lightweight tool which can be used for animating arbitrary BPMN models. It covers the most important basic sequence flow elements, and it can handle models of varying complexity. The tool can be used for introductory BPMN courses, for self-learning and experimenting with different process designs, but also for analyzing process models in industrial practice.

The remainder of the paper is structured as follows: In chapter 2, we discuss related work. The requirements for the tool are discussed in chapter 3. Chapter 4 describes the developed animation tool. Chapter 5 concludes the paper and gives an outlook on further work.

2   Related Work

The execution semantics of a process model can be analyzed by actually executing the model in a process engine. However, process engines are rather complex and often expensive, and for making a process executable, a lot of additional artifacts need to be developed, such as data structures and variables, rules and conditions, user interfaces, service calls etc. [9], [10]. This is only an option for analyzing models which are to be executed anyway, but not for BPMN models that are not meant to be executed. Besides that, not all process engines vendors have implemented the BPMN standard completely. In many cases the execution semantics deviates from the standard. Some BPMS vendors have included animation features into their systems, e.g. Inubit [11] and IYOPRO [12].

Another means to analyze process models is provided by simulation tools. The objective of a process simulation is to analyze the dynamic behavior in respect to load distribution, resource capacities, waiting queue lengths, and cycle times [13], [14],

[15]. This means that the entire system with many process instances is considered, but not the process logic of a single instance. For example, at an exclusive split it is not important which conditions are true, but only the percentage of instances in which a specific path is selected. Therefore it is not possible to analyze the actual flow logic of single instances. Some simulation tools also provide animation components, e. g. iGrafx Process [16] and L-SIM [17]. However, a simulation requires a lot of information, such as statistical distributions of process instantiations, probabilities for selecting specific paths, etc. All this would not be required for analyzing the execution semantics of a model.

We have already mentioned that there are various tools containing syntax and rules checks, e. g. there is a tool for checking the modeling rules defined by Bruce Silver [18]. There are also several tools for business process verification, e. g. [19], [20], [21]. Such tools may help discovering some of the problems that can also be detected by analyzing the token flow, but the modeler neither can see the token flow in the incorrect model, nor in the corrected model, and it cannot be detected that an otherwise correct process model simply describes something else than intended.

Recently, some BPMN modeling tool vendors have included animation features in their tools. An example is "Innovator for Business Analysts" from MID [22]. However, most of these animation features have some drawbacks. "Innovator for Business Analysts", for example, does not duplicate tokens at a splitting parallel gateway, but it lets the user decide which *one* of the parallel paths he wants to animate. Such an animation component is more a presentation and discussion aid rather than a correct token flow animation. The modeling component of IYOPRO provides a better implementation of the BPMN semantics, but the animation is restricted to one process instance at a time. In all commercial systems, the animation feature is tightly integrated into the system, and it is not possible to access or change the implementation of the animation logics. There are also several research prototypes that include some kind of BPMN token flow animation of BPMN, e.g. [23], [24]. However, these prototypes are usually focused on analyzing specific questions concerning the model semantics rather than providing a general-purpose animation component.

3   Requirements

The main purpose of the tool is to visually animate the token flow of arbitrary BPMN models for presenting and analyzing the model's execution semantics. The tool is aimed at BPMN trainers, learners, and active modelers. BPMN trainers should be able to prepare demonstration models for their courses. Such models can be used for explaining the various BPMN elements. Trainers can also ask questions concerning the behavior of a specific model and use the animation afterwards in order to validate or correct the assumptions of the course participants.

The tool can also be used for self-learning. The learners can create their own models and test their assumptions by animating and experimenting with different process designs. In this way, the tool supports a more interactive and therefore more efficient way of learning than by just reading a book.

Active BPMN modelers can use the tool for analyzing their models and ensuring that the execution semantics correctly represent the intended behavior of the process.

In order to fulfill these purposes for a broad target group, the following requirements have been identified:

- The tool should be lightweight and independent of a specific platform and of a specific modeling tool. Although the seamless integration into a modeling tool would provide more comfort to the modeler, it was decided to develop an independent animation tool, so that it is useful for many BPMN practitioners regardless of their favorite modeling tool.
- It should support the BPMN 2.0 standard, using the standard exchange format as input. The models will be created with an existing modeling tool, exported in the BPMN 2.0 XML format, and then imported into the animation tool, where it will be displayed and animated. It will not be possible to modify the model in the animation tool.
- The tool should be able to visualize the token flow of various BPMN models.
- It should support the most important sequence flow elements, namely untyped start and end events, terminating end events, tasks, sub-processes (collapsed and expanded), gateways (data-based exclusive, parallel, inclusive), conditional flows and default flows (both originating from gateways and from activities).

  This selection neither is exactly identical to the common core set as identified in [25], nor to the "descriptive process modeling subclass" as defined in the BPMN specification [1]. Since some aspects in these sets are rather different to handle by an animation tool (e. g. handling correlation information for message flows), we have focused on pure sequence flow-related elements, also including some elements from the analytical subclass of BPMN 2.0.
- Important BPMN elements that do not directly affect the token flow should also be displayed, namely pools, lanes, groups, and annotations.
- The tool should ensure correct token flow according to the BPMN specification.
- It should be possible to animate the tokens of multiple instances of the same process.
- It should be possible to interactively toggle the conditions of conditional flows (at gateways and activities) between *true* and *false*.
- Since the tool's purpose is the graphical animation, it is not required to support very large models, but only models of a size that fit onto a typical computer screen (with the element labels still readable).

4   BPMN Animation Tool

The BPMN animation tool has been implemented in Java. It is a standalone tool with a Java Swing user interface (Fig. 1).

For importing model files, the JAXP DOM interface is used. The model file is validated against the BPMN 2.0 XML schema. Unfortunately, the BPMN 2.0 standard exchange format is not used consistently by different BPMN tools. Therefore it was decided to use one modeling tool as the reference implementation. We selected the Signavio Process Editor [26] as reference tool, because this tool has a rather good implementation of the BPMN exchange format, and there is a free academic version available for research and teaching. This restriction to a reference tool means that we could not entirely meet the requirement of tool independence. However, this is due to the insufficient implementation of the BPMN standard by modeling tool vendors. By

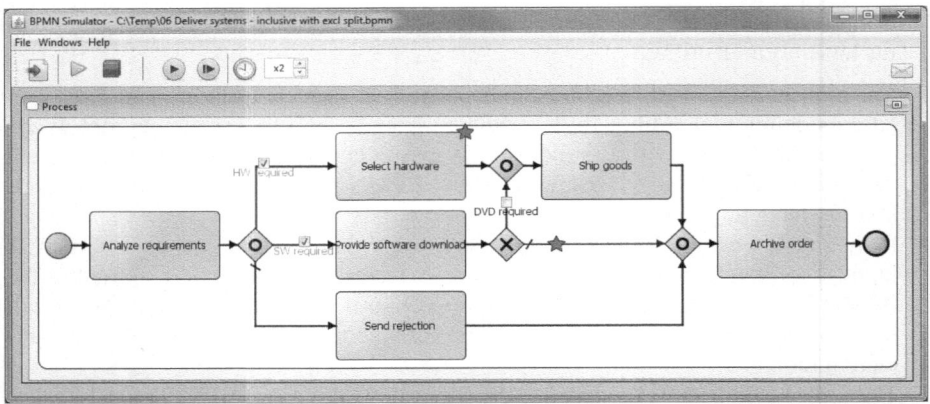

Fig. 1. User interface of the BPMN animation tool

using the standard exchange format, the animation tool is prepared to use the BPMN export from any tool that has implemented the standard correctly.

The internal structure is a rather straightforward implementation of those parts of the BPMN 2.0 meta-model that are relevant for the animation. The animation is handled by those BPMN elements that can contain or transport tokens. Each of these elements holds a list of its current tokens. At discrete time intervals it forwards tokens according to the BPMN execution semantics. The applicable execution rules are implemented in each element class.

Although the tool does not perform a complete syntax check on imported models, it checks for typical problems that prevent the model from being animated. For example, sometimes in the modeling tool a sequence flow connector is not really attached to the target object.

As Fig. 1 shows, the tokens are represented by little stars that move along the sequence flows. A process is instantiated by clicking on a start event or via the start button. When a task is activated, the star is shown on the task's upper right corner. A process can be instantiated multiple times. The tokens of different instances can be distinguished by different colors. Tokens of the same color are duplicated for parallel paths. Likewise, at parallel or inclusive joins, a token can only be joined with other tokens of the same color. At exclusive and inclusive splits there are tick boxes at the conditional sequence flows, so that it is possible to interactively change which conditions are *true*.

The animation can be paused and resumed, and the animation speed of the token flow can be changed. It is also possible to use a single step modus. The stop button removes all tokens from the model.

If one of several tokens with the same color reaches an end event, it is removed from the model. The number of removed tokens is shown in the upper right corner of the model, until the last token reaches an end event. When a token reaches a terminate end event, all tokens are immediately removed. At parallel or inclusive joins the waiting tokens of each color are displayed at the gateway. When all required tokens of the same color have arrived, they are removed and one resulting token is emitted at the outgoing sequence flow.

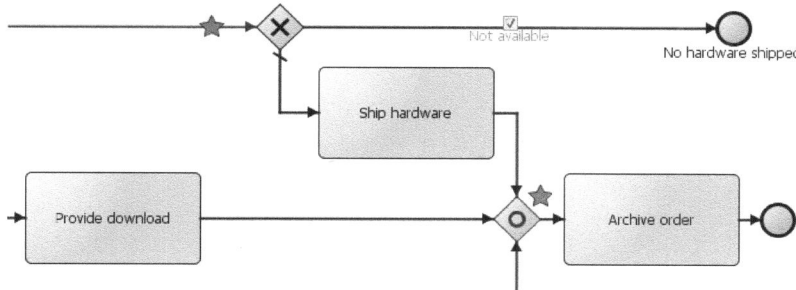

Fig. 2. The inclusive join waits for two tokens

The inclusive join has been implemented as defined in the BPMN specification. Thus, a situation as shown in Fig. 2 will be handled correctly. Here, the joining inclusive gateway waits for two tokens. One token has already arrived at the gateway.

When the other token moves to an exit of the exclusive gateway that leads to an end event, this token cannot reach the inclusive gateway any more. Therefore, the inclusive gateway does not wait any longer and immediately removes the single waiting token and forwards one token (Fig. 3).

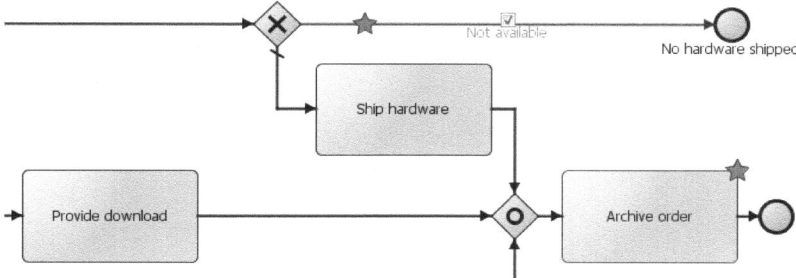

Fig. 3. The token at the top has left the path to the inclusive join, therefore the inclusive gateway has emitted a token. This token has activated the task "Archive order".

The tool can also handle the incorrect case that multiple conditions at an exclusive split are *true*. This can happen if the condition statements are defined in a way that a process variable may hold values for which more than one condition statement evaluates to *true*. For example, if the two condition statements are *"amount > € 5.000"* and *"amount ≤ € 6.000"*, both conditions are true for amounts between € 5.000 and € 6.000. Obviously, such a case is a mistake by the process designers. In the animation tool this – usually undesired – situation can be created by marking multiple exits of an exclusive gateway as *true* (Fig. 4).

According to the BPMN specification, the conditions at an exclusive gateway are evaluated one after the other. The first one that evaluates to *true*, determines which path will be taken. Since there is no prescribed order of the exits of a gateway, the selected path depends on the internal representation of these exits in a process engine – or in this case in the animation tool. This means that exactly one path with a *true* condition will be taken, but it is undefined which one.

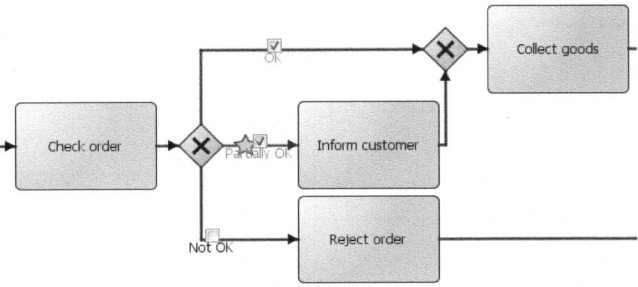

Fig. 4. The conditions of two alternative paths have been incorrectly marked as *true*. The tool routes the token to the first path it finds that is marked *true*.

If, on the other hand, no condition evaluates to *true* (and there is no default exit), the BPMN specification states that an exception will be thrown. In the animation tool, this incorrect situation is marked by a warning sign (Fig. 5). The process animation does not include explicit exception handling, but the user can handle this situation by either terminating the entire animation, or resetting the gateway exception state by marking one of the exits as *true*.

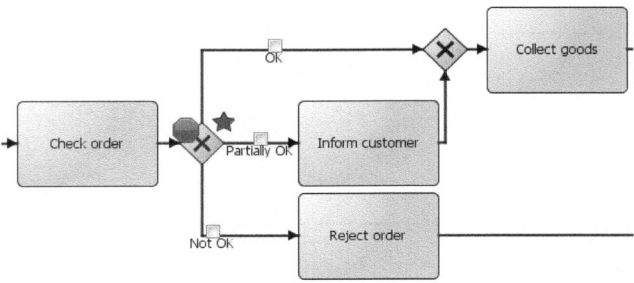

Fig. 5. The tool displays a warning sign, because no condition is marked as *true* and the process cannot continue

The tool can also animate the token flow in sub-processes, shown either in an expanded or collapsed view (Fig. 6). It is also possible to animate the token flow in a multi-level hierarchy of sub-processes.

For collapsed sub-processes, the detailed flow is shown in a separate window. While the sub-process is active, the token of the parent process is shown in the upper right corner of the collapsed sub-process. In the sub-process of Fig. 6, the parallel gateway emits two tokens. The token in the parent process does not continue its travel before both tokens have been consumed by the end events of the sub-process.

The tool has been released as Open Source under the Apache 2.0 license. It can be downloaded at http://code.google.com/p/bpmn-simulator. Instructions, sample processes, and videos can be found at http://www.kurze-prozesse.de/en/download-bpmn-token-flow-animation-tool.

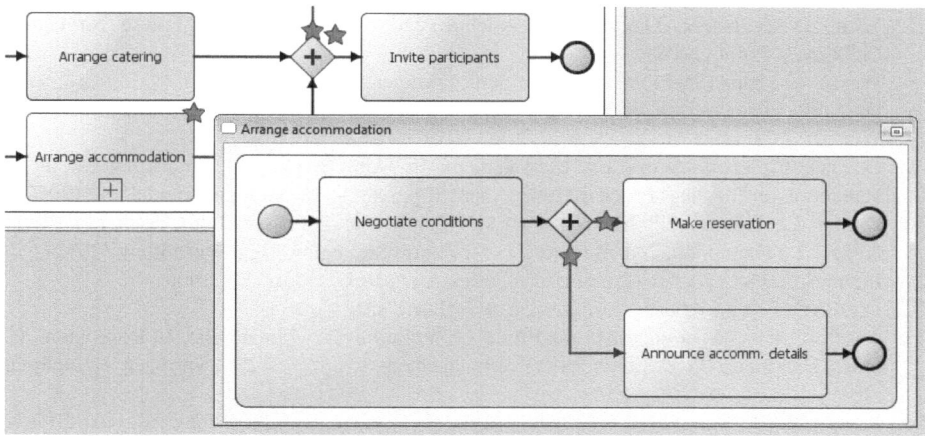

Fig. 6. Token flow in a collapsed sub-process. The detailed flow within the sub-process is shown in a separate window.

5   Conclusions and Outlook

The tool has been implemented successfully. A series of tests has shown that it can handle even rather complex models correctly. The objectives and requirements as stated in chapters 1 and 3 have been met. The only drawback was the insufficient implementation of the BPMN 2.0 exchange format by some tool vendors. Therefore we had to restrict the supported exchange format to the export format of one reference tool. However, since we use the standard format, it will not be difficult to extend the number of supported tools, as far as they fully support the standard.

Possible future extensions may include extended interaction features and further BPMN elements. For example, it could be useful to change the processing times of tasks, so that it is possible to change the order in which parallel tokens arrive at a gateway. Further supported BPMN elements could include intermediate events, attached events, and message flows.

The tool has been used successfully in several introductory BPMN courses. It could be useful to evaluate the benefits of the tool in different learning scenarios. For example, does it actually improve the learner's understanding of BPMN models, and does it help reducing the number of errors made by novice modelers?

References

1. OMG (ed.): Business Process Model and Notation (BPMN) Version 2.0. OMG document number: formal/2011-01-03 (2011), http://www.omg.org/spec/BPMN/2.0/PDF
2. Allweyer, T.: BPMN 2.0. Introduction to the Standard for Business Process Modeling. BoD, Norderstedt (2010)
3. Briol, P.: BPMN 2.0 Distilled. Lulu, Raleigh (2010)
4. Freund, J., Rücker, B.: Praxishandbuch BPMN 2.0, 3rd edn. Hanser, Munich Vienna (2012)

5. White, S.A., Miers, D.: BPMN Modeling and Reference Guide. Future Strategies, Lighthouse Point (2008)
6. Weske, M.: Business Process Management. Springer, Heidelberg (2007)
7. Dive Into Business Process Management, http://www.diveintobpm.org
8. Workflow Patterns, http://www.workflowpatterns.com
9. Ouyang, C., et al.: Workflow Management. In: vom Brocke, J., Rosemann, M. (eds.) Handbook on Business Process Management 1, pp. 387–418. Springer, Heidelberg (2010)
10. Dugan, L., Palmer, N.: Making a BPMN 2.0 Model Executable. In: Fischer, L. (ed.) BPMN 2.0 Handbook, 2nd edn., pp. 71–91. Future Strategies, Lighthouse Point (2012)
11. Inubit Suite, http://www.inubit.com/en/inubit-suite.html
12. IYOPRO, http://www.iyopro.com/?lang=EN
13. van der Aalst, W., et al.: Business Process Simulation. In: vom Brocke, J., Rosemann, M. (eds.) Handbook on Business Process Management 1, pp. 313–338. Springer, Heidelberg (2010)
14. Januszczak, J.: Simulation for Business Process Management. In: Fischer, L. (ed.) BPMN 2.0 Handbook, 2nd edn., pp. 135–150. Future Strategies, Lighthouse Point (2012)
15. Waller, A., Clark, M., Enston, L.: L-SIM: Simulating BPMN Diagrams with a Purpose Built Engine. In: Perrone, L.F., et al. (eds.) Proceedings of the 2006 Winter Simulation Conference, pp. 591–597. IEEE, Piscataway (2006), http://www.informs-sim.org/wsc06papers/073.pdf
16. iGrafx Process (2011), http://www.igrafx.com/products/process
17. L-SIM Server for Business Process Simulation, http://www.lanner.com/en/l-sim.cfm
18. Silver, B.: BPMN Method & Style, 2nd edn. Cody-Cassidy Press, Aptos (2011)
19. van Dongen, B.F., Jansen-Vullers, M., Verbeek, H.M.W.E., van der Aalst, W.M.P.: Verification of the SAP reference models using EPC reduction, state-space analysis, and invariants. Comput. Ind. 58(6), 578–601 (2007)
20. Mendling, J.: Empirical Studies in Process Model Verification. In: Jensen, K., van der Aalst, W.M.P. (eds.) Transactions on Petri Nets and Other Models of Concurrency II. LNCS, vol. 5460, pp. 208–224. Springer, Heidelberg (2009)
21. Fahland, D., Favre, C., Jobstmann, B., Koehler, J., Lohmann, N., Völzer, H., Wolf, K.: Instantaneous Soundness Checking of Industrial Business Process Models. In: Dayal, U., Eder, J., Koehler, J., Reijers, H.A. (eds.) BPM 2009. LNCS, vol. 5701, pp. 278–293. Springer, Heidelberg (2009)
22. Innovator for Business Analysts, http://www.mid.de/en/products/innovator-for-business-analysts
23. van Gorp, P., Dijkman, R.: BPMN 2.0 Execution Semantics Formalized as Graph Rewrite Rules: extended version. Beta Working Paper series 353. Eindhoven University of Technology (2011), http://cms.ieis.tue.nl/Beta/Files/WorkingPapers/wp_353.pdf
24. Sörensen, O.: Semantics of Joins in Cyclic BPMN Workflows. Diploma Thesis. Christian-Albrechts-University Kiel (2009), http://www.is.informatik.uni-kiel.de/~ove/research/papers/2009-OrJoin.pdf
25. zur Muehlen, M., Recker, J.: How Much Language Is Enough? Theoretical and Practical Use of the Business Process Modeling Notation. In: Bellahsène, Z., Léonard, M. (eds.) CAiSE 2008. LNCS, vol. 5074, pp. 465–479. Springer, Heidelberg (2008)
26. Signavio Process Editor, http://www.signavio.com

Towards SecureBPMN - Aligning BPMN with the Information Assurance and Security Domain

Yulia Cherdantseva[1], Jeremy Hilton[2], and Omer Rana[1]

[1] School of Computer Science and Informatics, Cardiff University, UK
{y.v.cherdantseva,o.f.rana}@cs.cardiff.ac.uk
[2] Department of Informatics and Systems Engineering, Cranfield University, UK
j.c.hilton@cranfield.ac.uk

Abstract. The participation of business experts in the elicitation and formulation of Information Assurance & Security (IAS) requirements is crucial. Although business experts have security-related knowledge, there is still no formalised business process modelling notation allowing them to express this knowledge in a clear, unambiguous manner. In this paper we outline the foundational basis for SecureBPMN - a graphical security modelling extension for the BPMN 2.0. We also align the BPMN with the IAS domain in order to identify points for the extension. SecureBPMN adopts a holistic approach to IAS and is designed to serve as a "communication bridge" between business and security experts.

Keywords: information security & assurance, BPMN, extension.

1 Introduction

The importance of Information Security (InfoSec) and Information Assurance (IA) has been escalating over the last several decades as a result of the growing reliance of organisations on Information and Communication Technology (ICT) and the recognition of information as a key business asset. During the last decade, we have observed an increasing tendency to perceive InfoSec as a business enabler and to recognise the importance of IA - a comprehensive and systematic management of InfoSec in a networked world [1]. In this paper we refer to the Information Assurance & Security (IAS) knowledge area, which incorporates the knowledge acquired by both InfoSec and IA [1]. The realm of IAS includes (1) all InfoSec countermeasures; and (2) a systematic and comprehensive management of these countermeasures. IAS is not limited to the technical aspect of information protection, it includes organisational, legal and human-oriented aspects as well.

Until recently, the IAS concerns were not considered at the stage of Business Process Modelling (BPM). Often, it is attributed to the fact that business experts have not enough security-related knowledge or training [2]. Nevertheless, the empirical studies show that business experts may express security needs at a high level of abstraction [3]. Business experts have knowledge essential for security design, e.g. knowledge about information levels of sensitivity, internal

J. Mendling and M. Weidlich (Eds.): BPMN 2012, LNBIP 125, pp. 107–115, 2012.

and external information sharing needs, and about legal and compliance IAS requirements (often sector-specific). Therefore, we see different reasons for the insufficient integration of security modelling into BPM. These are:

- the lack of commonly agreed understanding of the IAS domain;
- the complexity of articulating security requirements together with functional requirements;
- the communication gap between business and security experts (business experts express security needs at the very high level of abstraction, whereas security experts operate at the detailed technical level);
- little software-tool support for incorporating IAS aspects in BPM.

Overall, BPM is deemed to be a suitable foundation in order to fulfil the challenging tasks of security requirements elicitation and high level security design due to the following reasons:

1. the overall purpose of BPM is analysis and improvement of business processes in terms of time-effectiveness and efficiency through allowing easy identification of the problematic areas [6]. Hence, BPM could be used in a similar way to identify security-related problems in business processes.
2. The concept of business process has great importance for business experts [4,5]. Business experts do not need to familiarise themselves with a new technique to express security concerns.
3. BPM is also used by software developers to capture the initial requirements for the system design [4,5]. Thus, modelling of InfoSec within business process models allows parallel modelling of functional and non-functional security requirements.

Among a variety of modelling languages the authors have chosen the BPMN [7] as the basis for the extension, guided by the following considerations: (1) it is easily understood by all parties involved in system development - from business analysts to technical personnel [8]; (2) it supports modelling of collaborative business processes; and (3) it allows connection of business process design with implementation in a standardised way [7].

Contribution. In this research we aim to enrich the BPMN with the IAS modelling capabilities by developing SecureBPMN - a graphical security modelling extension for the BPMN 2.0. Here we outline the intermediate results of the SecureBPMN development project. This paper does not go as far as to present the finalised graphical notation, but discusses the need for and outlines the foundational basis of it. In Section 2, we give the overview of the related work. Section 3 outlines the concept behind SecureBPMN and the research method. Section 4 aligns BPMN with the IAS domain to show missing capabilities of the BPMN and points for the future extension. In Section 5, we draw conclusions and sketch a plan of further work.

2   Related Work

Over the last decade a number of research projects were conducted in an attempt to bridge the gap between the IAS and BPM domains. In 2009, the detailed survey of nine attempts to integrate security and risk aspects into business process management was presented by Jakoubi et al. [9]. Jakoubi et al. identified several gaps in the research. Our research aims to address two of them: (1) Extend a list of security goals and (2) Improve one of the business process modelling notations, namely the BPMN. With regards to the first point, we not only extend a set of security goals, but build a comprehensive model of IAS which, apart from security goals, includes information taxonomy, security mechanisms and stages of the IAS development life-cycle.

In 2007, Rodriguez et al. [2] proposed a BPMN extension that allows incorporation of security into BPM from the business analyst viewpoint. The authors of [2] develop a set of graphical concepts representing security semantics. Rodriguez et al. extend the Business Process Diagram (BPD) metamodel with five security requirements: Non-repudiation, Attack harm detection, Integrity, Privacy and Access control. Each security requirement may be specified only for a certain core element of a BPD and has a graphical representation - a padlock symbol with a corresponding capital letter in the center (Figure 2).

In 2008, Wolter et al. [10] discussed a model-driven transformation from security goals, specified in business process models in a graphical fashion, into concrete security implementations in the process-aware information systems, based on Service-Oriented Architecture (SOA). In this work a security concept is presented, which includes the following entities: object (a basic entity of the concept), security goal, constraint (fulfils a security goal), security mechanism (characterises techniques used to enforce a security constraint) and policy (defines constraints). Wolter et al. [10] use the existing BPMN *Group* element to depict security goals as well as a new element - security annotation - which consists of a graphical symbol and an accompanying text description (Figure 3).

In 2011, Mulle et al. [11] proposed a language for formulation of security constraints embedded in the BPMN. The authors address two gaps in the research: (1) incompleteness of security modelling vocabulary; (2) insufficient user involvement. The proposed language uses a standard BPMN *Artifact* element as a container for constraints. A constraint is presented as a structured text annotation. The main aim of the proposed language is to translate security requirements specified in a BPMN model into the executable specification. Hence, the language is text-based and oriented on technical experts. As a result, business experts find it hard to understand. This complicates the initial security requirements gathering.

In 2012, Saleem et al. [12] developed a Domain Specific Language (DSL), based on the BPMN. The proposed DSL allows modelling security requirements along the business process model in SOA applications. The BPMN metamodel is extended with essential security objectives. In comparison with [2], a limited set of security requirements is considered: Confidentiality, Integrity and Availability (associated with Non-repudiation). Saleem et al. also developed a set of graphical notations for Confidentiality, Integrity and Availability (Figure 4).

In 2012, Altuhhova et al. [13] conducted an analysis of the BPMN in terms of its suitability for security requirements derivation and expression of security

countermeasures. Altuhhova et al. [13] align the BPMN constructs with the
domain model of Information Systems Security Risk Management (ISSRM) [14]
and conclude that the BPMN requires an extension in order to be fully applicable
for security modelling.

Problem Statement and Solution Outline
The recent research has noticeably extended the existing body of knowledge and
advanced the area. Nevertheless, there are still some aspects that are not fully
addressed in the works discussed above and which SecureBPMN aims to address.

First, many authors still concentrate purely on the technical aspect of IAS
and do not address organisational, human and legal aspects. In order to address
this issue SecureBPMN adopts a holistic view on the IAS domain and takes into
consideration and allows modelling of security mechanisms of different natures.

Second, the research lacks an agreed, shared understanding of the IAS do-
main. This leads to the incomprehensiveness of a set of security goals being con-
sidered, and to the confusion between security goals and security mechanisms.
As a solution to this problem, we develop a solid theoretical IAS foundation for
SecureBPMN which is expressed via the ontology of the IAS domain and the
Multi-Dimensional Model of IAS (MMIAS) [15].

Third, the existing security extensions suffer from granularity. The research
considering the expression of security goals by business experts is isolated from
the research considering the selection of security mechanisms which help to
achieve those goals. Security modelling does not yet facilitate communication
between various experts (e.g. business, domain and security experts) involved
in the design of *secure* business processes and does not allow representation of
all security-related aspects in a consistent, traceable way. As a response, SecureBPMN aims to provide a notation that, first, allows consistent modelling
of all elements of the IAS domain and, second, enables modelling from different
viewpoints.

Fourth, none of the research discussed above performed an evaluation and
validation of the proposed security modelling notation with end-users to ensure
its clarity and practical applicability. SecureBPMN has two levels of validation:
(1) the IAS ontology and the MMIAS validation by InfoSec and IA experts;
(2) SecureBPMN notation validation by business process modelling and security
experts.

Fifth, the existing works aim to provide a way for incorporating some security
aspects into the BPMN, but omit the fact that the modeller may not have
sufficient or complete knowledge about the IAS domain. Foreseeing this issue,
we attempt to provide domain- and context-specific security recommendations
to a modeller during the process of security annotation.

3   The General Concept behind SecureBPMN and Research Method

SecureBPMN is a firm stepping-stone on the way to solve the problems identified
above. The general concept behind SecureBPMN is depicted in Figure 1. A

Business Process Model, which is annotated with security elements in line with the SecureBPMN semantic rules is referred to as a Secure Business Process Model (SBPM). Figure 1 shows that a Business Process Model is transformed into a SBPM by undergoing through the Assisted Security Annotation Process (ASAP), when an Expert annotates it with security elements.

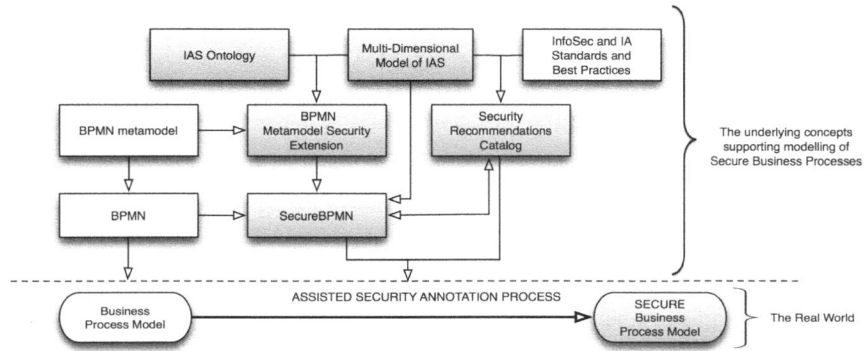

Fig. 1. The General Concept behind SecureBPMN

Above the dotted line, Figure 1 depicts the components required to enable the design of SBPM and the interrelationship between these components. The components shown in Figure 1, which are innovative and developed in this research project, are shaded. The un-shaded components illustrate the existing notations and concepts. Thus, the concept behind SecureBPMN includes the ASAP as well as:

- The IAS Ontology, which clarifies the interdependences between the fundamental elements of IAS, namely asset, security goal, security mechanism, threat, vulnerability and risk.
- The Multi-Dimensional Model of IAS (MMIAS) - a distilled, concise overview of the IAS domain, which has been developed on the basis of the analysis of the existing InfoSec and IA models. It fosters the commonly-shared domain understanding, reuse of the existing knowledge, makes ideas sharing easier, and promotes consistency of security policies and mechanisms across organisations.
- The BPMN metamodel extended with security entities and attributes outlined in the ontology and MMIAS;
- A Security Recommendations Catalog (SRC) - a database of the Security Recommendations, which is formed on the basis of the security-related standards and best practices and intended to assist a modeller, who has no in-depth knowledge of the IAS domain;
- SecureBPMN - the syntax, semantics and notation of the security modelling extension.

Table 1. Steps of the research method

Step Title	Input	Output
Mapping of the IAS knowledge area	InfoSec and IA academic and industry publications, standards, existing models of InfoSec and IA	IAS ontology; MMIAS [15]
Metamodelling	IAS ontology; MMIAS	Extended BPMN metamodel
Development of ASAP	IAS ontology; MMIAS; extended BPMN metamodel	ASAP
Development of SecureBPMN Graphical Notation	IAS ontology; MMIAS; extended BPMN metamodel	SecureBPMN
Development of a SRC	IAS ontology; MMIAS; extended BPMN metamodel; SecureBPMN	SRC
Prototype implementation	IAS ontology; MMIAS; extended BPMN metamodel; SecureBPMN; ASAP	Software tool supporting SecureBPMN

The research method which is used for the development of SecureBPMN consists of six steps outlined in Table 1 along with the expected output of each step. Although there is a required logical consequence of the steps, in practice the research and development of the extension is conducted in a spiral iterative, rather than a step-by-step manner.

4   Aligning the BPMN with the IAS Domain

The IAS ontology and the MMIAS, which are elaborated in this research project, form a grounded conceptual foundation of SecureBPMN. The detailed description of the ontology and MMIAS is given in [15]. The ontology and MMIAS define security elements and their attributes that are essential for the IAS domain and, therefore, should find their representation in the business process models. This section analyses how identified essential security elements and their attributes could be illustrated by the existing BPMN elements. Table 2, shows (1) correspondence between the IAS ontology elements and their attributes, and the MMIAS elements; (2) how elements of the IAS ontology and MMIAS may be represented by the BPMN elements; and (3) representation of security elements in the existing security extensions for the BPMN.

Table 2 shows that the majority of security elements could be represented with a *Text Annotation* BPMN element. Unfortunately, the usage of Text Annotation for expression of security elements of different nature is highly likely to lead to multiple misinterpretations of the security annotations in business process models. Although the security extension of the BPMN should fully use the existing BPMN elements, there is still a need to introduce new graphical elements for the visualisation of the following key security elements: security goal and its level of criticality; security mechanism; asset level of sensitivity; and access permissions for all actors within the model.

Thus, the analysis summarised in Table 2 confirms that (1) the BPMN syntax is insufficient for the representation of all elements outlined in the IAS ontology

Table 2. Alignment of the BPMN with the IAS ontology and the MMIAS

IAS Ontology Elements and their attributes	MMIAS elements	BPMN elements	Representation in other works
Security Goal	Security Goal	None; Possible: Text Annotation	[2] - padlock symbols (security requirements) (Figure 2); [10] - group element, text annotation with icon (Figure 3); [12] - colour symbols (security stereotypes) (Figure 4)
Criticality of Security Goal	Prioritisation of security goals	None Possible: Text Annotation	[2] - Security requirement has a level of criticality. No visual representation.
Asset	Information Taxonomy characterises the asset	DataObject, Message, DataStore	Present in the BPMN, no need for extension
Asset Sensitivity	Information Level of Sensitivity	None; Possible: Text Annotation	Not found
Asset State	Information State	Defined according to the position within a model	Not found
Asset Position/Location	Information Position/Location	Defined according to the position within a model	Not found
Security Mechanism	Security Mechanism	Activity, Task, Group, Association, Transaction, Compensation	[10] - group element, text annotation with icon (Figure 3); [11] - text annotation; [16] - blue circle with text description
Vulnerability	Not present	None; Possible: Text Annotation, Association, Message Flow, Task, Activity	[13] - Message Flow
Threat	Not present	None; Possible: Pool, Lane, Activity, Task, Message Flow	[13] - Message Flow, Text annotation, Pool, Lane
Risk	Reflected by the criticality of security goals	None; Possible: Text Annotation	[16] - red triangle with exclamation mark accompanied by text description
Access Permissions depend on Asset Sensitivity	Access Permissions depend on Information Level of Sensitivity	None; Possible: Text Annotation	[2] - a padlock symbol accompanied by text annotation (access permissions; security role);

Fig. 2. The Representation of Security Requirements by Rodriguez et al. [2]

and the MMIAS, and requires an extension to facilitate effective security modelling, and (2) currently, there is no comprehensive security modelling extension for the BPMN that allows clear, consistent representation of all elements of the IAS domain and their attributes.

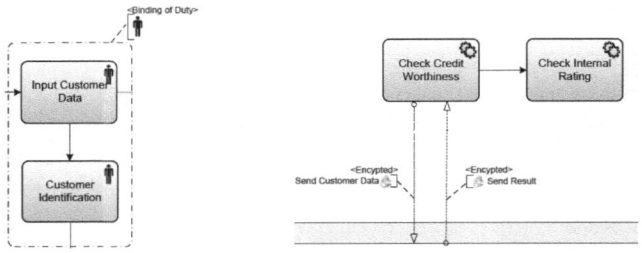

Fig. 3. The Representation of Security Goals by Wolter et al. [10]

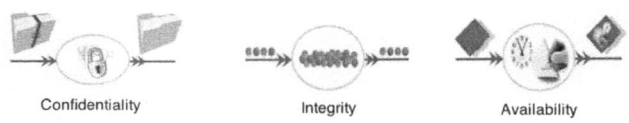

Confidentiality   Integrity   Availability

Fig. 4. The Representation of Security Stereotypes by Saleem et al. [12]

5   Conclusion and Further Work

This paper discusses the importance of consideration of IAS issues at the stage of BPM and presents the foundational basis for SecureBPMN - a security modelling extension for the BPMN 2.0. SecureBPMN will operate at the high level of abstraction and serve as a bridge between business and security experts. There are several important features that differentiate SecureBPMN and, as a result, determine its **novelty**: (1) a solid theoretical IAS foundation, (2) a holistic approach to IAS (modelling of technical, organisational, human-oriented and legal security mechanisms); (3) consistent modelling of all key IAS elements (and their attributes), and (4) support of security-decision-making process through the provision of security recommendations.

The research conducted so far allowed us to identify security elements that should be incorporated into the BPMN and to set the basis for the development of a visual notation. Further work involves the elaboration of the SecureBPMN graphical notation and its validation with end-users. The evaluation of the positive effect of the suggested extension will be carried out by applying SecureBPMN on a real-life case study and discussing the results with business and security experts.

References

1. Cherdantseva, Y., Hilton, J.: Information Security and Information Assurance. The Discussion about the Meaning, Scope and Goals (May 2012), http://users.cs.cf.ac.uk/Y.V.Cherdantseva/Cherdantseva_Hilton_2012.pdf (accessed on June 22, 2012)

2. Rodriguez, A., Fernandez-Medina, E., Piattini, M.: A BPMN Extension for the Modeling of Security Requirements in Business Processes. IEICE - Trans. Inf. Syst. E90-D, 745–752 (2007)
3. Lopez, J., Montenegro, J., Vivas, J., Okamoto, E., Dawson, E.: Specification and Design of Advanced Authentication and Authorization Services. Computer Standards and Interfaces 27(5), 467–478 (2005)
4. Leymann, F.: BPEL vs. BPMN 2.0: Should You Care? In: Mendling, J., Weidlich, M., Weske, M. (eds.) BPMN 2010. LNBIP, vol. 67, pp. 8–13. Springer, Heidelberg (2010)
5. Völzer, H.: An Overview of BPMN 2.0 and Its Potential Use. In: Mendling, J., Weidlich, M., Weske, M. (eds.) BPMN 2010. LNBIP, vol. 67, pp. 14–15. Springer, Heidelberg (2010)
6. Giaglis, G.: A taxonomy of business process modeling and information systems modeling techniques. International Journal of Flexible Manufacturing Systems 13(2), 209–228 (2001)
7. The OMG, Business Process Model and Notation (BPMN) Version 2.0 (January 03, 2011), http://www.omg.org/spec/BPMN/2.0 (accessed on June 22, 2012)
8. Wolter, C., Schaad, A.: Modeling of Task-Based Authorization Constraints in BPMN. In: Alonso, G., Dadam, P., Rosemann, M. (eds.) BPM 2007. LNCS, vol. 4714, pp. 64–79. Springer, Heidelberg (2007)
9. Jakoubi, S., Tjoa, S., Goluch, G., Quirchmayr, G.: A Survey of Scientific Approaches Considering the Integration of Security and Risk Aspects into Business Process Management. In: International Workshop on Database and Expert Systems Applications, pp. 127–132 (2009)
10. Wolter, C., Menzel, M., Meinel, C.: Modelling Security Goals in Business Processes. In: Proc. GI Modellierung, vol. 127, pp. 197–212 (2008)
11. Mulle, J., Stackelberg, S., Bohm, K.: A Security Language for BPMN Process Models. Karlsruhe Reports in Informatics (September 2011)
12. Saleem, M., Jaafar, J., Hassan, M.: A Domain-Specific Language for Modelling Security Objectives in a Business Process Models of SOA Applications. AISS 4(1), 353–362 (2012)
13. Altuhhova, O., Matulevicius, R., Ahmed, N.: Towards Definition of Secure Business Processes. In: WISSE 2012, Gdansk, Poland (June 2012), http://gsya.esi.uclm.es/WISSE2012/papers/paper5.pdf (accessed on June 27, 2012)
14. Mayer, N.: Model-based Management of Information System Security Risk. Doctoral Thesis, University of Namur (2009)
15. Cherdantseva, Y., Hilton, J., Rana, O.: SecureBPMN - a New Approach to Achieving Synergy between Information Security and Business Process Modelling (February 2012), http://users.cs.cf.ac.uk/Y.V.Cherdantseva/SecureBPMN.pdf (accessed on June 22, 2012)
16. BOC Group. Risk management and compliance with ADONIS: Community Edition, http://www.adonis-community.com/fileadmin/media/documents/RM_with_ADONISCE.pdf (accessed on May 21, 2012)

A BPMN Extension for Including Data Quality Requirements in Business Process Modeling

Alfonso Rodríguez[1], Angélica Caro[1], Cinzia Cappiello[2], and Ismael Caballero[3]

[1] Department of Computer Science and Information Technologies,
University of Bio Bio, Andrés Bello s/n, Chillán, Chile
{alfonso,mcaro}@ubiobio.cl
[2] Dipartimento di Elettronica e Informazione – Politecnico di Milano
Piazza Leonardo da Vinci 32, 20133 Milano, Italy
cappiell@elet.polimi.it
[3] Alarcos Research Group-Instituto de Tecnologías y Sistemas de la Información,
University of Castilla-La Mancha, Paseo de la Universidad 4, Ciudad Real, Spain
ismael.caballero@uclm.es

Abstract. BPMN is a notation for business process modeling through which it is possible to represent multiple characteristics of the analyzed business processes. However, although in a business process data play a fundamental role, it is still not possible to model data quality issues using BPMN due mainly to the lack of a specific notation. Since data quality is one of the main elements for achieving the business process goals, we aim to develop a comprehensive framework that supports the design of data quality-aware business processes. In this paper, we mainly focus on the part related to the elicitation and definition of data quality requirements and we present an extension of BPMN suitable to include them at a business process modeling level.

Keywords: Data Quality Requirements, Business Process Modeling, BPMN.

1 Introduction

Data Quality (DQ) is a critical component of the business activities, especially in information-intensive organizations. DQ is often defined as "fitness for use", i.e., the ability of a data collection to meet users' requirements [1]. DQ is usually evaluated by means of different dimensions (e.g., accuracy, completeness, timeliness and consistency), which selection mainly depends on the context of use. Poor DQ exposes organizations to non-depreciable risks. In fact, wrong or unreliable data might reduce the efficiency of the Business Processes (BP) and the effectiveness of the decisions [1]. These problems can be avoided by adopting suitable DQ assessment and improvement actions. Such actions often require the modification of the procedures that organizations use to perform the core business activities. Considering that the way in which the business tasks are connected is defined during the BP modeling phase, we claim that DQ issues should be already addressed at this stage.

For modeling BP, we refer to BPMN that has been adopted as a de facto standard [2]. Anyway, BPMN has been extended in various ways in order to represent different

J. Mendling and M. Weidlich (Eds.): BPMN 2012, LNBIP 125, pp. 116–125, 2012.

process characteristics. Since none of the extensions include DQ requirements in a BP model, we provide a framework to support all the activities related to the design of DQ-aware BP, ranging from the definition of DQ requirements to the identification of improvement actions able to guarantee the desired DQ levels. This paper briefly describes the whole framework but mainly focuses on *dqBP* (Data Quality in Business Process), a BPMN 2.0 extension for modeling DQ requirements.

The paper is organized as follows. Section 2 discusses the related work. The framework for the design of DQ-aware business processes is presented in Section 3. Section 4 provides details on *dqBP*. Section 5 shows the usefulness of the proposed approach by means of an example. Finally, our conclusions are drawn in Section 6.

2   Data Quality in Business Process Modeling

In spite of the benefits of managing data with an adequate level of DQ, business people should become aware of DQ requirements from the BP modeling phase. The most important languages used to model BP, BPMN and UML [2], do not allow process modelers to fully specify DQ requirements. BPMN has been already extended to consider different aspects: customer needs related to quality aspects such as time and cost [3], non-functional properties such as performance and reliability [4], temporal constraints [5], security requirements [6], and legal constraints [7]. However, there is only a specific notation, IP-MAP [8], to represent DQ issues in BP. In [9] authors made a first attempt to map IP-MAP constructs related to the DQ to BPMN but the limited diffusion of IP-MAP is a strong barrier for the use of such notation.

Anyway, DQ concerns are not new to the BP literature: some contributions highlight the need of addressing DQ issues in BP modeling. In [10], Soffer explores issues related to data inaccuracy. The work proposes an approach to design robust processes and avoid accuracy problems. Bringel et al. [11] propose a BP pattern to ensure DQ in an organization. The pattern consists in the definition of a BP model that can be reused through adaptation in different organizational scenarios. For this, they define DQ attributes associated with information entities that have different meanings on the basis of the analyzed business view. The Data Excellence Framework [12] describes the methodology, processes and roles required to generate the maximum business value from the improvement of BP using DQ and business rules. In this approach, DQ requirements are specified as business rules. Bagchi et al. [13] introduce a BP modeling framework for quantitative estimation and management of DQ in information systems. They propose to analyze the BP structure in order to estimate errors arising in transaction data and the impact of the error propagation on the performance indicators. Heravizadeh et al. [14] proposed the QoBP framework for capturing the quality dimensions of a process. The framework helps modelers in identifying quality of functions, of input and output objects, of human and non-human resources. In particular, they specify eleven DQ attributes for the input and output information objects. Finally, Lu et al. [15] propose an approach to consider compliance in the BP design, incorporating a set of control objectives in the BP allowing process designers to assess the compliance degree of their design and to be informed on the cost of non-compliance. A DQ aspect considered is the data integrity. The cited studies consider different DQ dimensions as summarized in Table 1.

Table 1. Data Quality Dimensions identified in BP modelling

Work ▼ \ DQ ▶ Dimension	Integrity	Accuracy	Uniqueness	Completeness	Non-Obsolescence	Consistency	Timeliness	Objectivity	Believability	Reputation	Accessibility	Security	Relevancy	Value-added	Amount of Data	Interpretability	Understandability	Concise Rep.	Consistent Rep.	Easy of Manipulation
Lu et al. (2000)	x																			
Soffer (2010)		x																		
Bringel et al. (2004)	x		x			x	x	x	x	x	x	x	x	x	x	x	x	x	x	x
el Abed (2011)	x	x	x	x	x	x														
Heravizadeh et al (2008)	x		x			x	x	x	x	x	x	x	x	x						

3   A Methodology to Design Data Quality-Aware Business Processes

The goal of our work is to include DQ aspects in BP modeling in order to avoid inefficiencies due to various data related errors. To reach such goal, we have defined a methodology that is composed of four (4) stages (see Figure 1).

Fig. 1. Methodology to model DQ aware business processes

The first stage, *Data Quality Aware Business Process Modeling*, at a BPMN Descriptive Level, starts with the enrichment of a BP model with *high-level DQ requirements*. Such requirements are graphically expressed by means of a specific annotation called *DQ Flags* that have to be associated with data-related BPMN elements. The activities in this stage are performed by the Business People/Analysts, who, by using the new annotation, are able to highlight the data elements for which DQ should be guaranteed for the success of BP. The output of this stage is a BPMN model enriched with a set of DQ Flags.

During the second stage, *Data Quality Requirements Specification*, these high level DQ requirements are refined. In details, Business Analysts and DQ Experts will collaborate in analysing, from a DQ point of view, the BP model generated in the previous stage. Workers review each DQ Flag in order to define *low level DQ requirements*, i.e., identify the DQ dimensions, relevant for the considered data-related BPMN element, and the corresponding level of importance (Low, Medium, High). As a result of this stage, documentation about the DQ Flags, the data-related BPMN elements and the low level DQ requirements should be generated.

In the third stage, *Business Process DQ Viewpoint Analysis and Improvement*, DQ Experts and the Business Designers analyse the low level DQ requirements in order to decide the most suitable way to accordingly improve the BPMN model. The improvement actions aim to modify the BP (e.g., insertion of new activities) in order to minimize the risk due to poor DQ. Risk is assessed by considering: a) the importance of each DQ Flag for the success of the BP; b) the probability of use of the data-related element associated with the DQ Flag; c) the DQ Flag Overhead defined as the ratio between the number of new activities added to tackle with DQ requirements and the total number of activities in the BP. The output of this stage is a BP model improved with new activities able to guarantee the satisfaction of the DQ requirements.

Finally, the fourth stage, named *Data Quality Use Case Diagrams Generation,* needs the involvement of DQ experts, and System Analysts for the generation of DQ Use Cases diagrams for each DQ Flag. Such set of use case diagrams will become available as requirements for the software development. The generation of use cases is based on a repository that contains the Standard Use Cases (one for each DQ dimension considered) and the DQ requirements associate with each DQ Flag in the BP.

4   dqBP: BPMN 2.0 Extension to Model DQ Requirements

BPMN considers notational elements grouped in five basic categories (Flow Objects, Data, Connecting Objects, Swimlanes and Artifacts), so that the reader of BPMN can easily recognize and understand a diagram [16]. Also, BPMN provides an extensibility mechanism that allows extending standard BPMN elements with additional attributes. Extension attributes must not contradict the semantics of any BPMN element. The new elements added to BPMN diagram should also have the basic look-and-feel so that diagrams should be easily understood by any modeler or viewer. BPMN 2.0 provides formal specifications for extending constructs using class diagrams for a metamodel representation. BPMN Extension consists of four different elements: *Extension, ExtensionDefinition, ExtensionAttributeDefinition,* and *ExtensionAttributeValue.* The core elements of an *Extension* are the *ExtensionDefinition* and *ExtensionAttributeDefinition.* The latter defines a list of attributes (name and type of the new attribute) that can be attached to any BPMN element. This allows BPMN adopters to integrate any metamodel into the BPMN metamodel and reuse existing model elements [16].

In BP, the data usage is an evident fact. Various elements of BPMN are used to represent not only data but also data flows, e.g., Data object or Message. However, aspects related to data quality for this kind of elements are not still available in the current version of BPMN.

This section shows *dqBP*, a BPMN 2.0 extension, which could be used by business people/analysts to model a BP including the specification of DQ requirements for each kind of data elements. Data-related BPMN elements susceptible to be associated with DQ requirements should be identified and marked by means of a set of flags, named DQ Flags. Therefore, DQ Flags will highlight the BP elements where the DQ may be necessary or relevant, according to the perspective of the Business People who are not necessary expert in DQ area. The DQ Flags are considering as a high level DQ requirement.

Figure 2, in grey color, illustrates (a) the BPMN 2.0 extension metamodel class diagram for our proposal, (b) the relationships between the extension and a set of

BPMN elements by means of the dqFlag class. dqFlag, the main class of the metamodel, is related to Data Objects, Message, Message flow, Conversation, Data Store, and Activity.

Because of the rich expressiveness of BPMN 2.0 to represent the semantic of a BP, we decided to associate a symbol to the dqFlag class. This symbol, composed by merging the letters D and Q (Ð), must be used to mark up with a DQ Flag any of the six BPMN elements to indicate that they are susceptible to be associated with a DQ requirement. In addition, a new graphical representation for each one of the combination of the new symbol and the one corresponding to the data-related BPMN elements is also introduced (see Table 2).

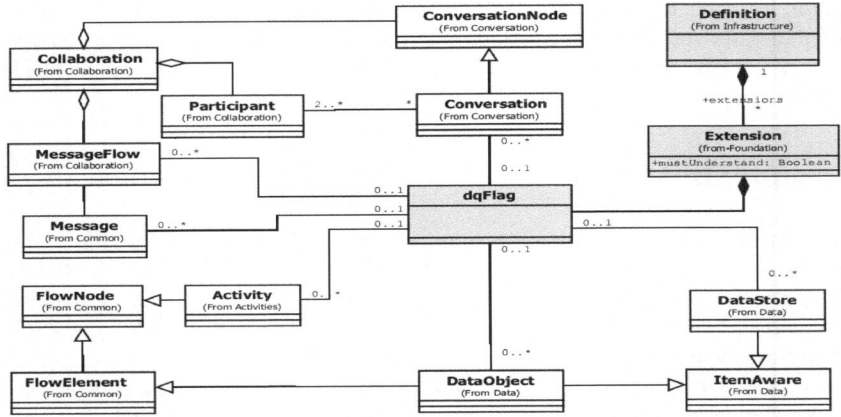

Fig. 2. BPMN 2.0 extension to support DQ Flags in the modeling of BP

Table 2. Representation of the combination of Data-related BPMN elements and DQ flags

Graphical Representation	Intended use of the Graphical Representation
Message	It represents that data contained in the message might satisfy some DQ requirements for the success the business process, e.g., Completeness and Consistency in a drug prescription from the doctor to a patient.
Message flow	It represents that data implicitly contained in the message might satisfy some DQ requirements, e.g., Currentness for a credit card authorization.
Conversation	It represents that data in some messages contained in the conversation might satisfy some DQ requirements, e.g., Security and Accuracy of the data interchanged between a customer and an airline Web application during the flight booking process.
Data Object	It represents that data in the data object might satisfy some DQ requirements, e.g., Completeness, Consistency and/or Accuracy of the data required to successfully deliver and ordered package to a customer.
Data Store	It represents that data contained in a data store might satisfy some DQ requirements, e.g., completeness of the data updated about product sale.
Activity	It represents that used/produced data in the activity might satisfy some DQ requirements, e.g., Precision and Accuracy of the budget generated as the output of one activity.

Figure 3 shows the metamodel to support the specification of low level DQ requirements starting from the analysis of a DQ Flag. In details, the class dqFlag-Specification is responsible for the realization of a DQ Flag. This class is linked to one or more data elements (dataElement class). Each data element represents the specific data associated with the data-related BPMN element annotated with DQ Flag (e.g., a check). Each data element is linked with one or more DQ requirement (dqRequirement class), and each DQ requirement is linked to a DQ dimension (dqDimension class) that will represent the DQ aspect necessary to obtain the required DQ level. Each DQ dimension can be linked to one or more measures (dqMeasure). Furthermore, each DQ dimension is linked to a standard use case diagram (ucSubject) that represents the set of actions that can be used to improve the DQ level considering the specific dimension. Finally, the standard use case diagram is associated with one or more specific use cases (ucDetail class) and with one or more actors involved in it (ucActor class). In Table 3, we show more details about each class of the metamodel.

Table 3. Classes of the metamodel for BPMN extension

dqFlag: Abstract class containing data quality flag specifications associated with a BP element in the BPMN model. Each data quality flag must be indicated in detail in dqFlag-Specification realization.   **Associations:**   Message[0..*];   MessageFlow[0..*];   Conversation[0..*];   DataObject[0..*];   DataStore[0..*]; Activity[0..*]   **Atributes:** None; **Realization**: dqFlag-Specialization; **Notation:** $\mathrm{I\!Q}$
dqFlag-Specialization: This class represents the DQ Flag associated with a BP element in the BPMN model.   **Associations:** dataElement [1..*]   **Attributes:** *bpFlag (int)*: Contains identification of element in BP where the symbol for DQ was indicated; *bpAc (str)*: Identification (properly the name) of activity directly related with BP element where DQ Flag specification was done. This value is empty when the flag is specified on Activity element; *bpPGE (int)*: Contains the number of previous exclusive gateways in relation with the activity previously identified. This value will be used to calculate the probability of execution for the activity; *bpItoBP (str)*: Contains a value that represents the importance of DQ to the business process success (low, medium, or high).
dataElement: This class represents the data elements (e.g., a check, an invoice, a report) related with a DQ Flag.   **Associations:** dqFlag[1..1]; dataRequirement[1..*]   **Attributes:** *deName (str)*: Short name with which the data element is known; *deDescription (str)*: Long description of the data element; **deSupport** *(str)*: Describes the data element support. Normally is possible distinguish between electronic or not electronic support; *deSource (str)*: Identify the part of the organization or external source where the data element is created or sent.
dqRequirement: This class represents the relation between data quality dimensions and a data element.   **Associations:** dataElement[1..1]; dqDimension[1..1]   **Attributes:** *dqrPriority (str)*: Contains a value high, medium, or low to specify the importance for this requirement in relation with other requirements for the same data element. This information is important take decisions when two or more requirements are in conflict; *dqrThreshold (str)*: Contains the minimum level required for the dimension associated with the DQ requirement of a data element.
dqDimension: This class represent the data quality dimensions associated with the DQ requirements.   **Associations:** dqRequirement [1..*]; dqMeasure [1..*]; ucSubject [1..1]   **Attributes:** *dqdDescription (str)*: Definition of the data quality dimension; *dqdAdAc (int)*: Number of activities necessary to implement the data quality dimension.
dqMeasure: Contain a basic identification of measure related with a data quality dimension.   **Associations:** dqDimension[1..1]
ucSubject: Contains an identification of standard use case related with data quality dimensions.   **Associations:** dqDimension[1..1]; ucDetail[1..*]; ucActor[1..*]
ucDetail: Contains a description of the all use case related with the data quality dimension.   **Associations:** ucSubject[1..1]; ucDetail[0..*]; ucActor[1..*]
ucActor: Contains the actors identification related with the use case.   **Associations:** ucSubject[1..1]; ucDetail[1..*]

Using this BPMN extension, our methodology will allow the design of business processes together with the DQ requirements able to guaranteeing the required data quality level needed to avoid data-related errors and to achieve BP objectives.

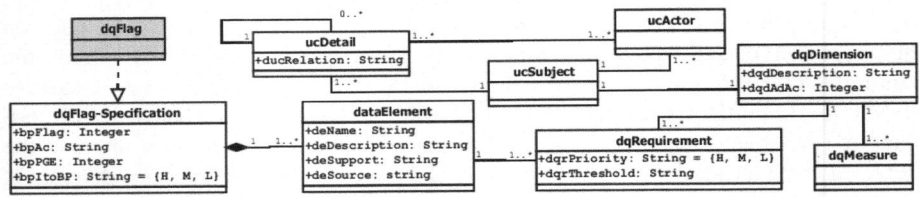

Fig. 3. Metamodel specification of low level DQ requirements in a BPMN model

5   An Illustrative Example

Let us consider the process related to the last phases of any product order management: payment and delivery of the ordered products. In our example the BP starts with the payment phase: the payment can be processed in two different ways: by credit card or by cash (or check). If the payment is made by credit card, it is necessary to ask for card authorization to the «Financial Institution». If the payment is performed by cash (or check), no controls are needed. When the payment is complete, the Distribution office prepares the package and delivers it to the customer.

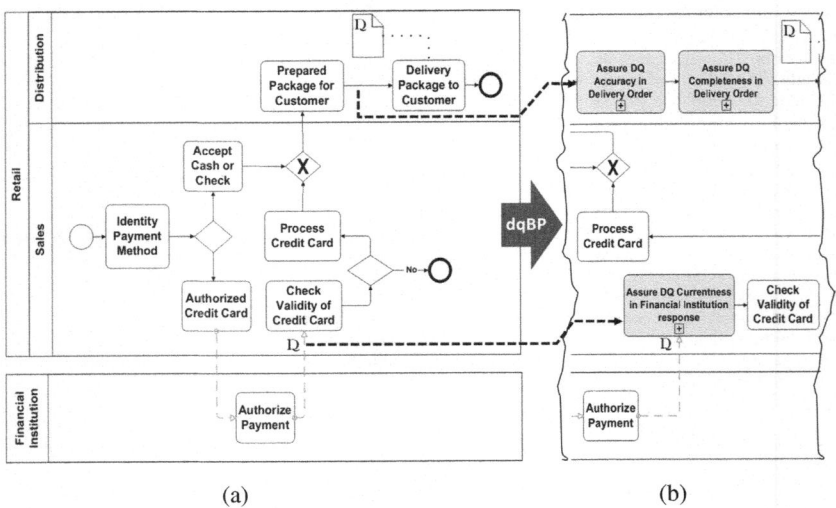

(a)   (b)

Fig. 4. Illustrative example: BPMN model with DQ Flags and BP improved

In order to enrich the described BP considering DQ Requirements we applied our methodology. In *"Data Quality Aware Business Process Modeling"* stage, business people have defined two DQ Flags. The first one, named *DQFlag1*, is associated with the DataObject needed as input in the *"Delivery package to customer"* activity. This DataObject contains the customer information necessary to deliver the package (identification, address). The second one DQ flag, named *DQFlag2*, is associated with the Message Flow (authorization or the rejection) coming from the *Financial Institution* pool to *Sales* lane. Figure 4 (part a) shows how the extension dqBP has been used: the data-related BPMN elements have been marked with the symbol ᴆ.

In *"Data Quality Requirements Specification"* stage, the workers reviewed each DQ Flag to specify the corresponding low level DQ requirements. Then, the definition of a *DQFlagSpecification1* and *DQFlagSpecification2* are required. The data elements on which DQ requirements have to be defined are: (i) *"Delivery Order"* and (ii) *"Financial Institution Response"*. See details about both DQ flags in Table 4.

In *"Business Process DQ Viewpoint Analysis and Improvement"* stage, the BP designer and DQ Expert decide which DQ improvement actions should be adopted. The suitability of each action is evaluated by considering: the importance of DQ Flag, the probability of execution and the DQ Flag overhead. In our example, *DQFlag1* has a High impact on the success of the BP. We assume that we do not have initial knowledge about the BP execution and starting from analysis of BP flow (considering the exclusive gateways), we have estimated that the probability of execution of the delivery action is 75%. The overhead associated with DQFlag1 is 25% because to tackle the DQ requirements two collapsed activities (remain the same grained used in BPMN descriptive level) for Accuracy and Completeness have to be included.

Table 4. DQ Flags specifications

(a)		(b)	
DQFlag1 → DQFlagSpecification1		DQFlag2 → DQFlagSpecification2	
BPMN element: Data Object	**P. exec.**: 75%	**BPMN element**: Message Flow	**P. exec.**: 50%
Influence: High	**Overhead**: 25%	**Influence**: Medium	**Overhead**: 12,5%
Name: Delivery Order	**Support**: Electronic	**Name**: Financial institution response	**Support** Electronic
Description: Delivery order (customer information)	**Source** Internal	**Description**: Delivery order (customer information)	**Source** Internal
DQ Requirements Accuracy (High) and Completeness (Medium)		**DQ Requirements** Currentness (High)	

DQFlag2 has a Medium impact on the success of the BP. The probability of requesting the payment authorization is 50% because when the payment is not performed by credit card the activity related with the DQ Flag is not executed. The overhead associated with this DQFlag is 12.5% because to tackle the DQ requirements one new collapsed activity must be included in the process. All new collapsed activities are shown in Figure 4 part b (in grey color), and each collapsed activity is shown in detail in Figure 5.

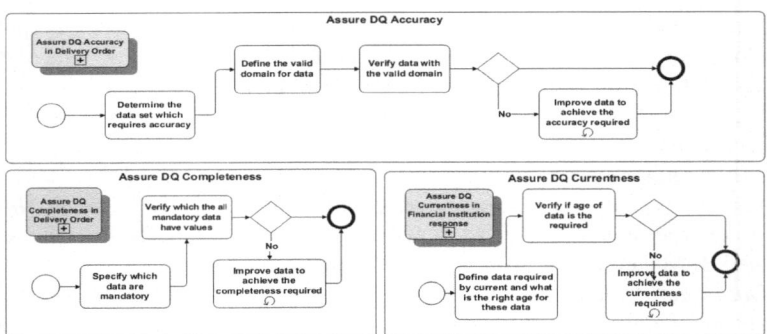

Fig. 5. BP model improved

Finally, *Data Quality Use Case Diagrams Generation* stage must generate the Use Case diagrams, which specify the DQ requirements for the software that will implement the improved BP model. Figure 6 shows for example the use case diagram generated for the *DQFlag2*. The use case diagrams delivered in this stage are general, but constitute a first approach towards the software development.

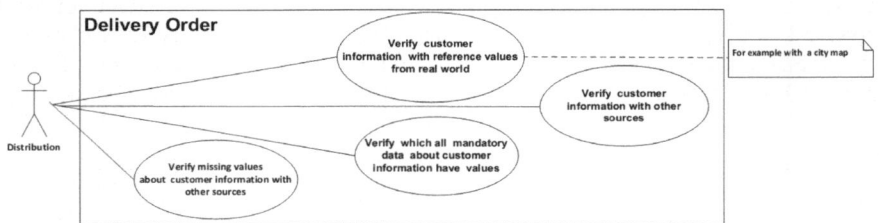

Fig. 6. Use case diagram for the BP model

6   Conclusions and Future Work

In this paper, we have presented a BPMN 2.0 extension which allows business people/analysts to specify DQ requirements in BP models. In details, we propose to include a special symbol to mark the different data-related BPMN elements in order to show the importance of DQ for the success of the business process. This extension firstly allows business people to be aware of the DQ issues (BPMN descriptive level). Secondly, it supports the improvement of the BP model including new activities able to tackle DQ (BPMN analytic level). The extension is used in a framework that includes a methodology that is composed of four stages and provides a systematic way for DQ management.

Our future work will focus on the refinement of the stages of methodology. Moreover, we will create a tool for automatize the described framework. Finally, real case studies will be considered in order to test the efficiency and effectiveness of our proposal.

Acknowledgments. Projects: MECESUP (UBB0704), and IQMNet (TIN2010-09809-E).

References

1. Wang, R., Strong, D.: Beyond accuracy: What data quality means to data consumers. Journal of Management Information Systems 12, 5–33 (1996)
2. Harmon, P., Wolf, C.: Business Process Modeling Survey. Business Process Trends (2011), http://www.bptrends.com/
3. Saeedi, K., Zhao, L., Falcone Sampaio, P.R.: Extending BPMN for Supporting Customer-Facing Service Quality Requirements. In: Proceedings of ICWS, pp. 616–623 (2010)
4. Bocciarelli, P., D'Ambrogio, A.: A BPMN extension for modeling non functional properties of business processes. In: Proceedings of TMS, pp. 160–168 (2011)
5. Gagne, D., Trudel, A.: Time-BPMN. In: Proceedings of CCEC, pp. 361–367 (2009)
6. Rodríguez, A., Fernández-Medina, E., Piattini, M.: A bpmn extension for the modeling of security requirements in business processes. IEICE Transactions E90(D-4), 745–752 (2007)
7. Goldner, S., Papproth, A.: Extending the BPMN Syntax for Requirements Management. In: Dijkman, R., Hofstetter, J., Koehler, J. (eds.) BPMN 2011. LNBIP, vol. 95, pp. 142–147. Springer, Heidelberg (2011)
8. Shankaranarayanan, G., Wang, R.Y., Ziad, M.: Ip-map: Representing the manufacture of an information product. In: Proceedings of ICIQ, pp. 1–16 (2000)
9. Sánchez-Serrano, N., Caballero, I., García, F.: Extending BPMN to Support the Modeling of Data Quality Issues. In: Proceedings of ICIQ, pp. 46–60 (2009)
10. Soffer, P.: Mirror, Mirror on the Wall, Can I Count on You at All? Exploring Data Inaccuracy in Business Processes. In: Bider, I., Halpin, T., Krogstie, J., Nurcan, S., Proper, E., Schmidt, R., Ukor, R. (eds.) BPMDS 2010 and EMMSAD 2010. LNBIP, vol. 50, pp. 14–25. Springer, Heidelberg (2010)
11. Bringel, H., Caetano, A., Tribolet, J.: Business Process Modeling Towards Data Quality Assurance. In: Proceedings of ICEIS, pp. 565–568 (2004)
12. el Abed, W.: Data Governance: A Business Value-Driven Approach (2009)
13. Bagchi, S., Bai, X., Kalagnanam, J.: Data quality management using business process modeling, pp. 398–405. IEEE (2006)
14. Heravizadeh, M., Mendling, J., Rosemann, M.: Dimensions of business processes quality (QoBP), pp. 80–91. Springer (2009)
15. Lu, R., Sadiq, S., Governatori, G.: On managing business processes variants. Data & Knowledge Engineering 68, 642–664 (2009)
16. Object Management Group: Business Process Model and Notation, BPMN 2.0 (2011)

Author Index